POPULATIONS IN A Seasonal Environment

MONOGRAPHS IN POPULATION BIOLOGY

EDITED BY ROBERT H. MACARTHUR

1. The Theory of Island Biogeography, by Robert H. MacArthur and Edward O. Wilson
2. Evolution in Changing Environments: Some Theoretical Explorations, by Richard Levins
3. Adaptive Geometry of Trees, by Henry S. Horn
4. Theoretical Aspects of Population Genetics, by Motoo Kimura and Tomoko Ohta
5. Populations in a Seasonal Environment, by Stephen D. Fretwell

POPULATIONS IN A Seasonal Environment

STEPHEN D. FRETWELL

PRINCETON, NEW JERSEY
PRINCETON UNIVERSITY PRESS
1972

LC Card: 75-173759
ISBN: 0-691-08105-0 (hardcover edn.)
ISBN: 0-691-08106-9 (paperback edn.)
This book has been composed in Linofilm Baskerville
Printed in the United States of America

TO MR. HAINES

Preface

The Subject

Most organisms live in a seasonal environment. In the course of a life cycle, some species face cold winters and hot summers, seasons of little rain and seasons of abundant rain, seasons of migration and seasons of more stable residence. Many have breeding and nonbreeding seasons. The numbers of individuals in a species at the beginning of any one of these seasons is dependent on the survivors from the preceding season. The population will grow in size during the course of a breeding season, but will decline during the course of a nonbreeding season. During seasons of population growth, the population may not expand sufficiently to fill all the available habitat. During seasons of decline, parts of the population may temporarily exist in habitats generally unsuitable to the species.

We face the general question: What are the factors regulating the distribution and abundance of species populations? As species populations vary in either distribution or abundance with the seasons, so must the answer to this question vary. The population during one season will, of course, be primarily influenced by factors operating during that season. It will also be influenced by factors operating at other seasons. The interaction of a population with a regularly varying environment is thus a complex problem which merits detailed analysis. I shall attempt some of this analysis in the following chapters, considering first the regulation of numbers (population size) in a seasonal environment (Section I). In Section II, I shall consider the habitat distribution of populations.

Philosophy—Strategy

A discussion of philosophy is, or should be, a plea for tolerance, not gratuitous advice. Whatever works should be acceptable, but in order to facilitate the making of value judgments, we tend to construct artificial standards which let us easily and quickly prejudge the value of things. In a changing world, however, the facility of prejudice becomes its only virtue. Thus, we tend to reject the unfamiliar, even when it may have real merit.

The philosophy of science which I follow throughout is that which seems most acceptable to those who study the efficiency of discovery. Philosophers of science take the scientists' word as to which discoveries are of value. The philosophers then attempt to formalize the procedural steps that evidently worked so well. The philosophy often recommended is the hypothetico-deductive (H-D) philosophy (Tricker, 1965; Ghiselin, 1966), which most of us learned in high school.

The basic steps in this philosophy are: (1) summarize experience (published, personal, or both) in an explanation; (2) formalize the explanation into a hypothesis; (3) deduce a testable prediction from the hypothesis; (4) test the prediction with data (new experience); (5) if data match prediction, go to step (3), otherwise go to step (1).

Most biologists will find this statement familiar. Many may even teach it to their students. Yet there are aspects of this philosophy that are not so familiar. I shall mention three of the most important.

1. According to the H-D philosophy, there is no such thing as unwarranted speculation. Speculation (guesswork, hypothesis formation) can be legitimate before any data have been gathered. This is less true of data collection, however. Unwarranted (theory-free) data gathering has little place in H-D science. This is, in fact, one of the major

weaknesses of this particular philosophy of science, for theory-free data can be productive.

2. H-D science is cyclic, and so there can be no conclusions made about hypotheses. The remark "The data are not sufficient to support the conclusions (about hypotheses)" does not have any meaning in H-D science. In the H-D philosophy, support of hypotheses is a matter of verification of predictions, but hypotheses can never be absolutely proved. Sufficiency of data is related to the verification of predictions and is entirely defined by statistical significance, if at all.

3. In H-D science, there is little virtue in considering alternative explanations. There are, it seems, an infinite number of these alternate explanations for any set of data points, so rejecting alternates (and doing nothing else) is an impossibly long task. Rejecting only "simple" alternates might be feasible, if simplicity were well defined. But there is no stated advantage in considering alternate explanations in the H-D philosophy until the hypothesis under consideration either fails to generate new predictions or fails to have its predictions verified. Tricker (1965) provides a proof that repeated verifications of predictions from a single hypothesis are sufficient to establish the "truth" of the hypothesis.

The main difference between hypothetico-deductive and more descriptive philosophies lies in one's attitude towards uncertainty. All of the three points just noted indicate that the H-D scientist is never quite certain where he stands relative to the real world. He does not "believe" that a particular theory is or is not true and does not talk much about facts, scientific or otherwise. Ideas play an important role in his investigations, and any sets of definitions or models that permit a clearer grasp of particular ideas are valuable research tools.

The reader should not construe these remarks as an attack on more descriptive philosophies. I am personally of the opinion that really good scientists are basically amateurs, operating as their pleasure dictates and quite immune to philosophical advice. So I offer none to others, nor do I make any judgments about the proceedings of others. I hope the reader is similarly inclined and will use the above remarks primarily to improve his understanding of their author.

ON THE STRATEGY OF BEING AN ECOLOGIST

Introduction

This section was originally developed as a teaching tool, to explain Levins' fitness set ideas (1962) to mathematically uncertain students.

The idea was to give the terms "fitness set" and "adaptive function" meaning by using them in discussing a human problem with which we are all very familiar. Ideally, interest in the problem would provide motivation to drive the student through the symbolic notation.

Also, the section provided an example of hypothetico-deductive science, using this methodology in an area where everybody is an expert. The inadequacies of these models *which are always present in any set of models* are conspicuous in this example. The oversimplifications and narrow assumptions can be recognized by almost anyone. This is not always so easy. We can easily be deceived by a complicated mathematical structure, assigning it more importance than it really deserves.

I offer this section here because of its value in setting the tone of the rest of the book. I have not provided the total answer to the problem of strategy in becoming an ecologist, but have captured only one probably-significant factor, or concept, in a well defined language. Any ecologist can now put that concept to use in his own decision-making process.

Levins' fitness sets provide a handle with which the idea can be grasped. The same sort of statement can be made throughout the book. I offer models not as my version of reality but as part (I hope) of the process that is ecological understanding.

Strategy

Ecology is traditionally nontheoretical or descriptive. Within the last half century, however, there has arisen some interest in theoretical approaches in ecology. Except in the case of statistical analysis, however, descriptive and theoretical ecology have remained largely separate. Even statistics need not qualify as an exception, as it is usually used descriptively. Adapting oneself to this dichotomy and to the use of the H-D scientific method poses an interesting problem.

This problem can best be described using the method of fitness sets (Levins, 1962). We consider the area defined by the axes: X_1, fitness in the presence (judgment) of theoretical ecologist (habitat 1); and X_2, fitness in the presence (judgment) of descriptive ecologist (habitat 2). These axes are equivalently: competence in theoretical ecology (X_1) and competence in descriptive ecology or data gathering (X_2) (referred to hereafter as "field" competence). We shall presume that theoreticians do not judge data-gathering ability, and vice versa.

In order to qualify for favorable judgment by either descriptive or theoretical ecologists, one must devote oneself to a period of study and independent (at least apparently independent) research. This requires time and does result in field or mathematical competence. However, time is limited and the student who strives for competence in theory must perforce neglect field development, and again, vice versa. Thus the strategies open to a student for a given amount of time expended result in some level of competence in theory (X_1) and some level of competence in data collection (X_2), but X_1 and X_2 are inversely related, so higher values of theoretical competence imply lower values of data competence.

There are two kinds of possibilities described in Figures 1 and 2. Figure 1 suggests that there is really a limited

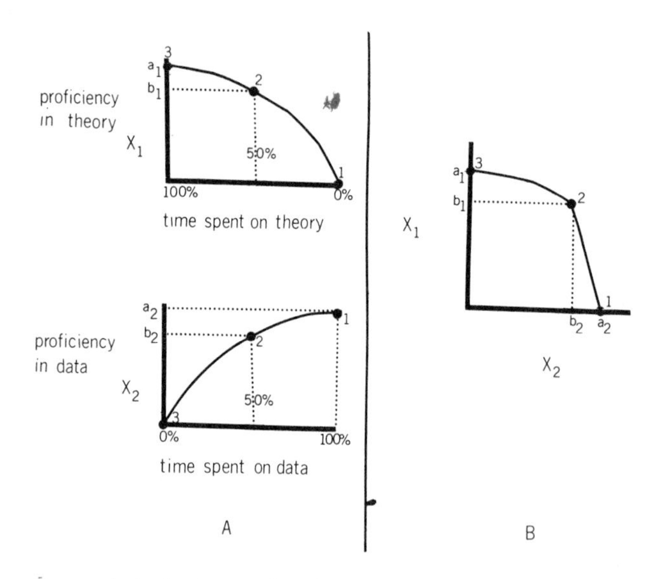

FIGURE 1. In the A part of the figure, proficiency is plotted against percentage of time invested. (The time axes are in opposite directions because more time spent on data means less time spent on theory.) The upper portion of A yields values for theory (X_1), the lower part, for data (X_2). For example, at 100% time spent in data (dotted line, points 1) data proficiency is a_2, 0% time is spent in theory and so no (0) proficiency is achieved there. For each different percentage of time, there is specified a particular proficiency in X_1 and a particular proficiency in X_2. These values are found by drawing vertical lines up from the percentage axes, at the percentage value desired. The dotted line is one example, the dashed line another. These X_1 and X_2 values can be plotted against one another, as in the right hand part of the figure (B). At $t_{x_2} = 100\%$ ($t_{x_1} = 0\%$) $X_1 = 0$ and $X_2 = a_2$. These values yield point 1. At $t_{x_2} = 50\%$ ($t_{x_1} = 50\%$), $X_1 = b_1$ and $X_2 = b_2$; these values yield point 2. When all time is spent in theory ($t_{r_1} = 100\%$) and none in data ($t_{r_2} = 0\%$), theory proficiency (X_1) is a_1, and data proficiency (X_2) is 0 (point 3).

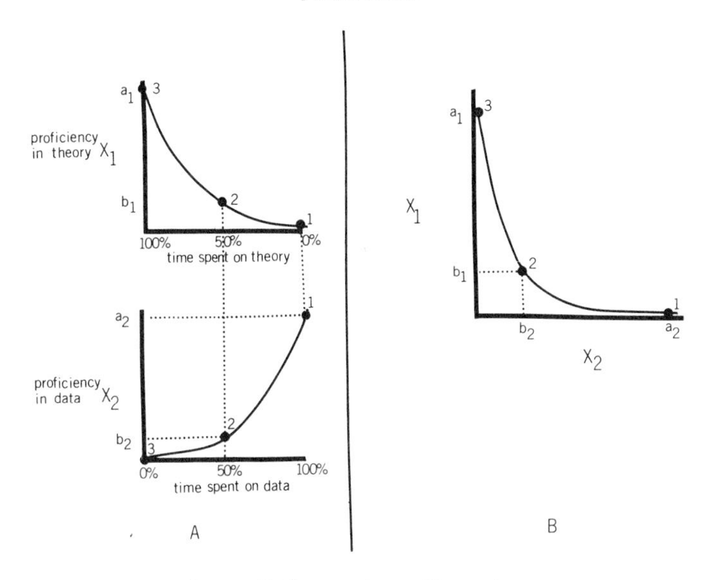

FIGURE 2. See caption to Figure 1.

amount of theory and data technique to learn, and that one can easily learn most of either area. By putting half of one's time in theory and half in data one masters most of both fields.

This possibility is probably not applicable to most ecology. Theoretical competence expands rapidly as new mathematical techniques are added to one's repertoire, so doubling the number of techniques at one's disposal more than doubles one's competence. The same is true in descriptive work where describing twice as many aspects of a system more than doubles one's understanding of what is going on in the system. Thus the student who devotes half of his time to theory and half to data probably knows much less than half as much theory or technique as the specialist student who concentrates on just one or the other. The relationship between X_1 and X_2 is probably as in Figure 2. The B parts of Figures 1 and 2 are "fitness sets."

What do we use for an adaptive function (measure of success) in developing the strategy of being an ecologist? There are two alternatives: We can measure success in terms of either prestige or progress. Prestige comes from the judgments of our peers and superiors. The average ecologist receives such judgments in encounters at meetings, in reactions to publications, and through other social or professional media. A proportion (P) of these encounters are with theoreticians and the rest ($1 - P$) are with descriptive scientists. The proportion will vary with the individual, and with particular situations. Assuming that one has theory competence X_1, and data competence X_2, that one receives prestige in exact measure to one's competence, and that one's theoretical competence means nothing to a descriptive ecologist (and vice versa), then

$$\text{Prestige} = Pr = PX_1 + (1 - P)X_2 \tag{1}$$

We have, however, in Figure 2B, the available strategies of X_1 and X_2 that a single person can achieve. These can be inserted in (1) to yield different prestige levels, for a given time expenditure. Levins found the highest success level in a given situation (i.e. fixed P) graphically. Using this approach, we fix prestige, Pr, and describe the various combinations of X_1 and X_2 that give this fixed value of prestige in the given environment (P) (Figure 3). We try different Pr values, picking the one that is achievable using a possible combination of X_1 and X_2 from the fitness set and is highest. Graphing Figures 2 and 3 together, we find (Figure 4) *achievable Pr* values by choosing those prestige lines that intersect the fitness set line. The highest Pr line of those achievable is easily seen to be the one that intersects the fitness set at $X_1 = a_1$ and $X_2 = 0$. If P were changed so that theoreticians were more frequently encountered, the prestige lines would shift in slope, and the maximum prestige would occur at $X_2 = a_2$, and $X_1 = 0$. In either case, it

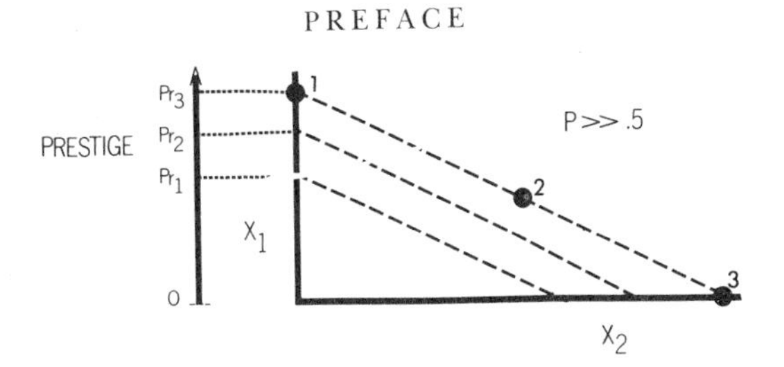

FIGURE 3. Prestige at various combinations of X_1 and X_2. At point 1, the scientist is gathering all his praise from theoreticians, and he is sustaining a prestige level of Pr_3 (the highest shown). In 2, he has given up half his competence in theory to gain competence in data. However, to stay at Pr_3, he has had to gain more competence in data than he gave up in theory (the horizontal distance of the point from the axes is greater than the vertical distance). The reason for this is $P \gg .5$; more encounters are with theoreticians and so data competence is less rewarding. At point 3, the scientist gets all his praise from data gatherers, and has put all his time into competence in field work. Since theoreticians are encountered more frequently ($P \gg .5$), a higher level of data competence is needed to keep at Pr_3. There would be no advantage to point 3, unless data competence were easier to attain. If $p < .5$, the lines would change in slope, being much steeper, and would intersect the X_1 axis at a higher value than the X_2 axis, the reverse of the present case. Pr_1 and Pr_2 are lower prestige levels.

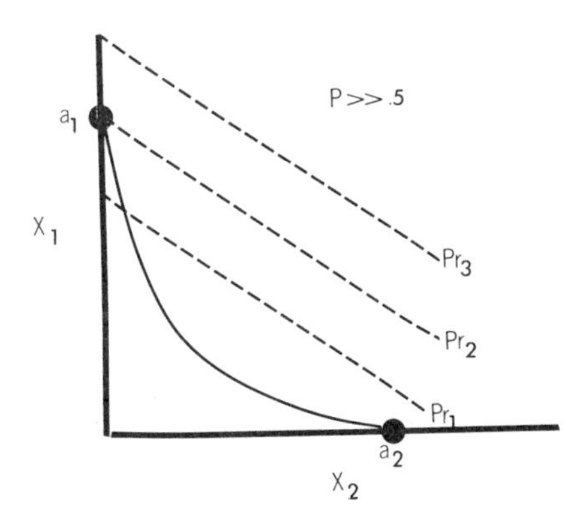

FIGURE 4. Prestige lines (adaptive functions, broken lines) plotted with the fitness set (solid line). The highest prestige attainable is Pr_2, which is achieved if the student studies only theory and attains $X_1 = a_1$, leaving $X_2 = 0$.

behooves one to be a specialist, if praise is what is desired.

Of course, insofar as a researcher habitat-selects, that is, ignores "the others," this model does not fit.

Scientific progress, our other adaptive function, is measured in terms which depend on scientific methodology. If, in fact, the hypothetico-deductive method is accepted as the best algorithm to use, then the adaptive function is different from the linear one given above. In gathering prestige, a theoretical snub does not reduce the number of compliments one gets from descriptive people. So the prestige from the one camp could simply be added to the prestige from the other camp. But in hypothetico-deductive (H-D) science this is not the case.

The steps in the H-D algorithm are, in general:

(1) Speculation
(2) Formal hypothesis formation (model building)
(3) Deduction-prediction
(4) Data gathering
(5) Data-hypothesis evaluation
(6) = (1) Explanation-speculation (if data refute prediction)
or (6) = (3) New prediction (if data verify prediction)

Steps (2) and (3) require theoretical competence, steps (4) and (5), technique. Each step depends on the ones preceding *and* following it. The theory is usually dependent on the data that are being explained and predicted; the data always depend on the theory that has predicted, or will explain it.

In describing the adaptive function for this case, we can surmise that an individual who is weak in theory but strong in the field will do a brilliant job collecting data that neither test a theory nor can be very elegantly explained. He will make some, but very little, H-D progress. Similarly, the person weak in laboratory or field work but firmly grounded

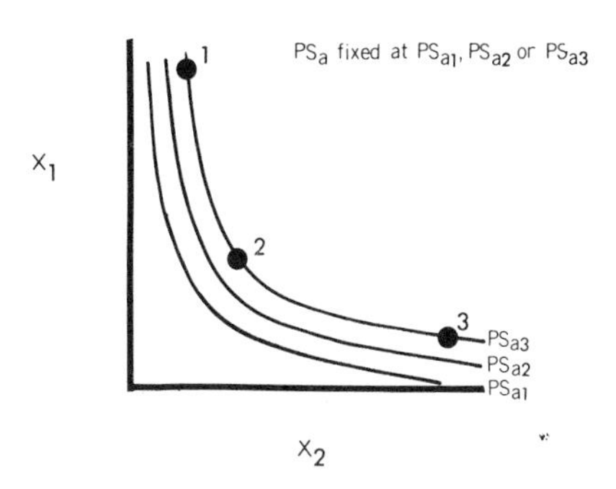

FIGURE 5. Hypothetico-deductive progress at various levels of X_1 and X_2. A particular rate of H-D progress (e.g., PS_{a3}) can be attained by having a great deal of theoretical competence, and relatively little data competence (point 1), a great deal of data competence but relatively little theory (point 3), or a moderate amount of both theory and data competence (point 2). This curve never intersects the axes, as no H-D progress can be made by either pure theory or pure observation.

in theory will offer beautiful theories that can explain only a very small part of the available data and that are almost impossible to test. He will achieve no more progress than the field man above. The same level of progress (denoted PS_a) achieved by the two specialists could be achieved by a person moderately competent in both theory and field work. Such a researcher would be able to explain much of the data he collected and would be able to test most aspects of whatever theory he could develop. Since the field man has data he cannot use and the theory man models he cannot apply or test, their extra competence in these areas is not efficiently used.

The three points described above are plotted in Figure 5, and the suggested adaptive function is drawn accordingly. Proceeding, using Levins' graphical techniques, we plot these adaptive functions together with the fitness set (Figure 6). We then consider only those adaptive functions that are achievable, that intersect the fitness set, and pick the one of these which represents the highest rate of progress.

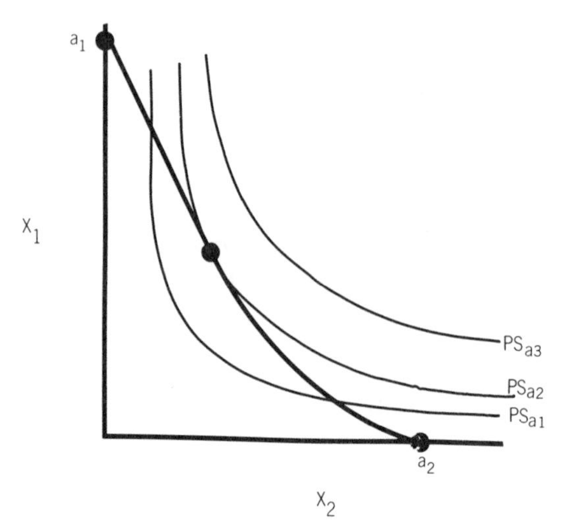

FIGURE 6. Progress adaptive function plotted with the fitness set. The heavy line is the fitness set, possible combinations of X_1 and X_2; the finer lines are the adaptive functions, from Figure 5. PS_{a2} is the highest rate of H-D progress that intersects the fitness set, that is, that has at least one attainable combination of X_1 and X_2.

The point of intersection is again the best strategy: In this case, this point implies that a scientist should mix his training, and become moderately competent in both theory and data gathering.

So there are two strategies available: specialized training and mixed training. The first optimizes prestige, the second progress. One can satisfy his ego or his curiosity, but not both. The mixed-strategy scientist may be criticized (justly!) for incompetence by both pure theorists and pure data biologists, as he proceeds to make satisfying advances in the science. The specialist will be frustrated by drawers full of unpublished data or untested theories.

Of course, I have not considered the effects of prestige that mixed-strategy biologists accord one another. This is because these scientists are sufficiently rare (in ecological circles at least) that their proportion is negligible. However, if ecology were to develop into a hypothetico-deductive science, perhaps this proportion would increase, and the decision as to strategy would not be so ambiguous.

SYNOPSIS

The central idea of this monograph is that the complexities of a seasonal environment significantly affect regulation of populations. Neglect of seasonal factors, where they occur, is apt to lead to a misunderstanding of how a population is interacting with its environment. This idea is developed essentially by example. I build simple (and very particular) models of populations in a seasonal environment (Chapters 1 and 2), deriving a few definitions and statements that reflect the surprising ways in which seasons affect population. These models are then applied to certain specific populations (Chapter 3), demonstrating the reality of seasonal regulation. For the most part, the first three chapters present and demonstrate a technique of population analysis that accommodates seasonal factors. However, they also support the idea that the biology of an organism during some season or seasons may have little importance in the determination of abundance.

Section II considers how a species population behaves during a particular season. The arguments in this section frequently depend on the population being limited during one or another season, and they are mostly concerned with distribution. The theories are not of very general applicability, being developed primarily with bird populations in mind. Again, I offer techniques of analysis for the understanding of particular species distributions. This section also points to the roles of experience and dominance in regulating animal distributions. Seasonality enters into these relationships in its effect on the movements (migration, dispersion, or changes in habitat distribution) of individuals in the population. The section also (Chapter 7) provides a general theoretical statement that population analysis is based on physiological ecology, adaptive morphology, and distribution of resources. This statement is developed in part in an example wherein other theoretical arguments are used to relate the quality and abundance of resources to physiological parameters, and to relate the location and distribution of resources to adaptive morphology.

I have at all points worked around examples, lest I stray too far toward the more intangible measures of worthiness. Besides solidifying the theory, this has resulted in considerable natural lore, especially about birds. I never approached my bird studies as an ornithologist, which has advantages and disadvantages. The primary disadvantage is the considerable possibility that natural facts exist and are known which will make conspicuous the naïveté of my own analysis. Fortunately, the hypothetico-deductive philosophy that I use is quite generous toward committing errors, and, as far as my reading in ornithology has gone, I feel that there is a small contribution to the understanding of bird populations in the examples I present here. Most of the studies or analyses here support the proposition that the nonbreeding season is of primary importance in the regulation of numbers of birds. Territorial or dominance behavior, or both, are thought to be important in some season in the dispersion of most bird species. Effects of experience on dominance may be indirectly important in all seasons. Theories of the distribution and abundance of breeding open nesting passerines and wintering fringillids are given. In the first theory, nest predation and experience are given a somewhat new emphasis. In the second, the distributions of food sizes and locations are of evident importance. These analyses offer answers to questions relating to why a passerine bird species lives where it does, or indicate where such answers will probably be found.

THE WRITING—ACKNOWLEDGMENTS

The ideas underlying the following chapters were roughly developed at North Carolina State University while I worked there as a graduate student in biomathematics (NIH Grant, GM-678) from 1964 to 1968.

None of the ideas or techniques are particularly novel; Carl Helms first stressed to me the importance of seasonality in avian biology, and Tom Quay, the importance of winter bird population studies. Drs. H. L. Lucas, Jr. and H. R. van de Vaart stimulated me to give serious attention to the philosophical aspects of model building in biology and encouraged my first attempts at it. Many of the modeling techniques I use were borrowed from MacArthur and Wilson's marvelous book on island biogeography (1967). MacArthur's work has been an invaluable source of stimulation and focus throughout. In the end, it was he who suggested that I put these ideas together in a monograph. Most of the monograph was outlined while I was on a Ford Foundation postdoctoral year at Princeton under MacArthur, 1968–1969. Much of the actual writing and much of the final analysis of data were completed while I was working with the Kansas State University biology faculty, whose patience with my preoccupation was exceeded only by that of my wife. This work was supported by NSF GB-14293. It is a pleasure to thank the Biology Division Secretarial Staff of Kansas State University, especially Karen Dungey, for quick, careful typing of several drafts of the manuscript.

Gordon Orians and Michael Rosenzweig read the complete manuscript, offering invaluable suggestions and criticisms. Various chapters were read by H. L. Lucas, H. R.

van der Vaart, S. Charles Kendeigh, Elliot Tramer, and Mary Wilson. The material has been used in the classroom and discussed with students. Gilbert Blankespoor, Ronald Case, and Melvin Taylor made especially useful observations and criticisms.

The development of most of the ideas was greatly helped by a rich and stimulating correspondence with M. Rosenzweig, and by blackboard discussions with H. Horn, G. R. Marzolf, and J. Zimmerman.

I wish to thank a number of agents for permission to use certain figures. Specifically, these are: Frederick Smith and the Duke University Press for Figure 25; H. G. Andrewartha and The University of Chicago Press for Figure 27; the British Ecological Society for Figures 30–33; *Acta Biotheoretica,* published by E. J. Brill, for Figure 37; and *Bird Banding* and the Northeastern Bird Banding Association for Figures 38–40 and Figure 42.

Contents

SECTION I

THE SIZE OF SPECIES POPULATIONS

CHAPTER ONE

Models for Long-Generation Species

Populations living in a seasonal environment are exposed to regular or systematic changes in resource quality and abundance. The populations respond to these fluctuations in resources with changes in population size and with qualitative changes in the way resources are used. In temperate regions, many populations grow during the summer and decline through the winter. One kind of food is consumed through the summer, another through the winter. One kind of predator affects the summer population, another kind affects the winter population. Some organisms are active in the summer but spend the winter hibernating, or in a diapause, or in a resistant egg. Many organisms migrate to new environments or habitats. Social systems sometimes change with the seasons.

This chapter offers some simple mathematical or graphical ways of analyzing seasonal trends in population size. The methods have been developed in order to answer the question: How will changes in the environment influence the population? I think that the answer to this question involves a thorough understanding of the population, its size and its fluctuations.

The size of populations living in a seasonal environment usually varies in a systematic fashion. The population increases during one or several seasons, may be stable during other seasons, and declines during the rest. On the average the declines equal the gains, so at a given point in the (annual) cycle of seasons the population shows only random changes in size. On the basis of this simple idea, we shall develop some ideas about populations in a seasonal environment. We shall, for the most part, restrict our attention to a simple alternation of seasons: a season of growth followed by one of decline, and shall consider two kinds of populations: one composed of organisms with generation time about equal to the seasonal cycle, "long-generation time," and (Chapter 2) one composed of organisms with generation time much shorter than the seasonal cycle. We shall also consider briefly organisms that have several seasonally-related stages in the life cycle. These particular kinds of theories are aimed at the kinds of observations generally available in the literature. They are intended to be more useful than complete. They are all deterministic, and this constitutes a major weakness. Hopefully, stochastic extensions, such as those presented by Pielou (1969) will be available to provide more accurate models.

POPULATIONS WITH LONG GENERATION TIMES

Assumptions About the Species

We shall proceed to describe a hypothetical species, making the following assumptions. The annual cycle is divided into breeding and nonbreeding seasons. In the breeding season adults produce offspring which become full grown (but not mature) by the beginning of the winter season. During the nonbreeding season, the population is reduced because of mortality of some of the adults and many or most of the new immatures. At the beginning of the next breeding season, all surviving individuals are adults which again try to breed. On the average, in a stable population, the number of adults at the beginning of each breeding season is constant.

We define production rate (b) as the average production of young per individual that survive to the beginning of the nonbreeding season minus the proportion of adults lost during the breeding season. Death rate (d) is the average percentage of adults and young failing to survive through the nonbreeding or winter season to the beginning of the breeding season. We shall assume that, for a fixed environment, production rate is always positive and decreases only with an increase in the initial size of the breeding population; and that death rate increases only with an increase in the initial winter population size. These assumptions are made to simplify the problem; therefore, the following theory may be severely limited in application. The assumptions are reasonable however, since many of the factors affecting birth and death rates are density-dependent and in a fixed environment population size is proportional to density.

The Stable Population—a Deterministic Model

The preceding description of a hypothetical species is sufficiently exact to permit the formulation of a mathematical model of population size. The model must be deterministic, specifying the population size only for the average case. It thus will determine only an average about which the real population size will vary.

Let the population at the beginning of the breeding season be of size N. At the end of the breeding season, bN individuals have been added (immatures added minus adults lost). The population is of size $N + bN$, or factoring, $N(1 + b)$. During the course of the winter season, $dN(1 + b)$ are lost and $N(1 + b) - dN(1 + b)$ survive. Thus the population becomes $N(1 + b)(1 - d)$. If the population is stable, this equals N, so

$$1 - d = \frac{1}{1 + b},$$

$$d = \frac{b}{1 + b}.$$

This suggests that we define replacement rate, g, as $b/(1+b)$, and we shall do so. Because of our assumption on b, the production rate, the replacement rate always decreases with an increase in population size. Also, $b/(1+b)$ or g is bounded by zero and one, as of course, is death rate, d. Note that these "rates," g and d, are defined for a *season* of population growth or decline and are *not* instantaneous. This theory provides no consideration of mid-season phenomena; only the lowest and highest population levels enter into the analysis.

The winter population size is initially $N(1+b)$ and we have assumed that death rate increases with an increase in this quantity. Any pair $(N,N(1+b))$ for which $b/(1+b)=d$ represents an equilibrium or balanced population. The breeding replacement exactly balances the winter (or nonbreeding) mortality. These population sizes $(N,N(1+b))$ are the deterministic averages about which the population sizes (spring and fall) might vary. At any other values of

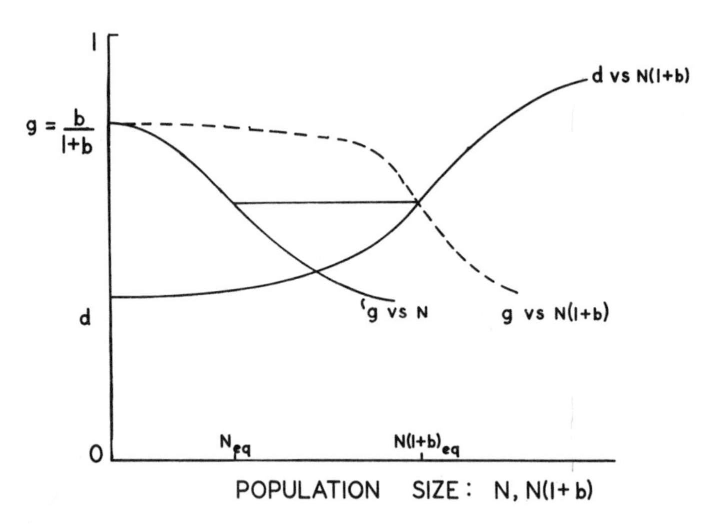

FIGURE 7. Graph of model of population regulation in a seasonal environment.

N,$N(1 + b)$, the replacement rate would not balance the mortality and the populations would either decrease or increase.

Figure 7 graphically presents this model. Two solid curves are drawn: g (intersects $N = 0$ at upper left) and d (intersects $N = 0$ at lower left). The sigmoid nature of the curves in Figure 7 is an assumption based on resource exploitation curves proposed by Peterson, Lucas, and Mott (1965) and on density-dependent predation curves observed by Tinbergen et al. (1960) and me (Chapter 6, Figure 44). The abscissa is population size: N for the g curve, $N(1 + b)$ for the d curve.

Transformation Curves

Also plotted (dashed line) is the transformation of the g curve onto the $N(1 + b)$ coordinate system. We shall use this transformation curve as a means of finding the stable equilibria.

The transformation curve is the fall population expressed as a function of the preceding season's reproductive rate. It is computed by horizontally adding bN to the reproductive rate curve. The intersection of this transformation curve with the d curve determines the equilibrium value of $N(1 + b)$. The horizontal difference between the g curve and the g transformation is, by definition, Nb, the population added in the breeding season at breeding population size N. Therefore, where the transformation curve intersects the d curve, the horizontal distance between the d curve and the g curve also is bN. On a horizontal line however, $d = b/(1 + b)$, so the population *lost* during the nonbreeding season is $dN(1 + b) = [b/(1 + b)][N(1 + b)] = bN$. Since the loss, bN, equals the gain (also bN), N,$N(1 + b)$ represents an equilibrium pair.

If g decreases sharply enough with N, the transformation curve folds back and has the form of a backward S. Then

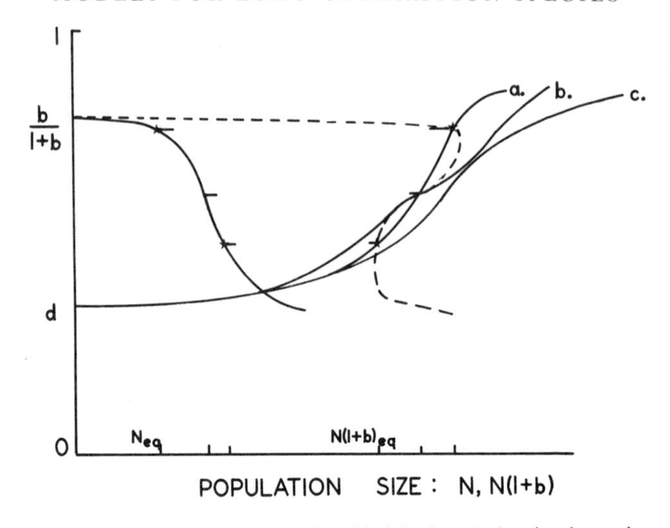

FIGURE 8. Population control when birth is sharply density-dependent (see text).

(curve a of Figure 8) the *d* curve may intersect it at three points, or (curve b of Figure 8) the *d* curve may coincide with it over a range of points, or (curve c of Figure 8) only one intersection may occur. However, if three intersections should occur, only two are stable equilibria. The center equilibrium in the region where the transformation curve is folded back is such that at slightly greater sizes of breeding populations, *g* decreases, but *d* decreases (i.e., increases negatively) even more, so the replacement rate becomes higher than the death rate. The breeding population then will increase to the next higher equilibrium. At breeding population sizes slightly lower than equilibrium, death rate is greater than replacement rate and the breeding population will decline to the next lower equilibrium. Only the edge equilibria are stable, the center equilibrium simply being the point of transition from one state to the other. A population with considerable random variation in population size and a folded transformation curve with three intersections might be expected to vary about one equilibrium for a period then jump to the other and vary there for a period. Takasaki (1964) has proposed similar theoretical populations via an alternative analysis.

Of course, the replacement and death rate curves are symmetric. The entire theory could be worked out with an N transformation curve of the death rate. Triple intersections occur only when the transformations of both the d and the g curves are simultaneously folded. In both cases, the decrease or increase in the rates with small changes in population size must be considerable (see Appendix to this chapter). Also, the transformation curves may fold and refold and any odd number of intersections may occur, but only every other one is stable. There is, however, no point in extending this aspect of the theory further until some of the curves have actually been estimated.

The Effect of Variation in Resources of Only One Season

Let us see, in terms of the model, what would occur if there were to be an increase in the area of more suitable habitat during just one season. If the new habitats were utilized, there would be a reduction of density in all the habitats occupied by the population with a concomitant relaxation of density-dependent effects. For example, if the habitats are all increased by a fraction, a, then the density in each habitat is decreased (on the average) by $1/(1 + a)$. This would change the replacement rate, which is density-dependent. A population size increase of $(100 \times a)$ percent would restore all the densities and therefore the replacement or death rate. Thus a season's curve is moved to the right (to higher population sizes) exactly $(100 \times a)$ percent when its habitat is increased $(100 \times a)$ percent. This means that at zero population size there is no shift $[0 \times (1 + a) = 0]$; at very large population sizes

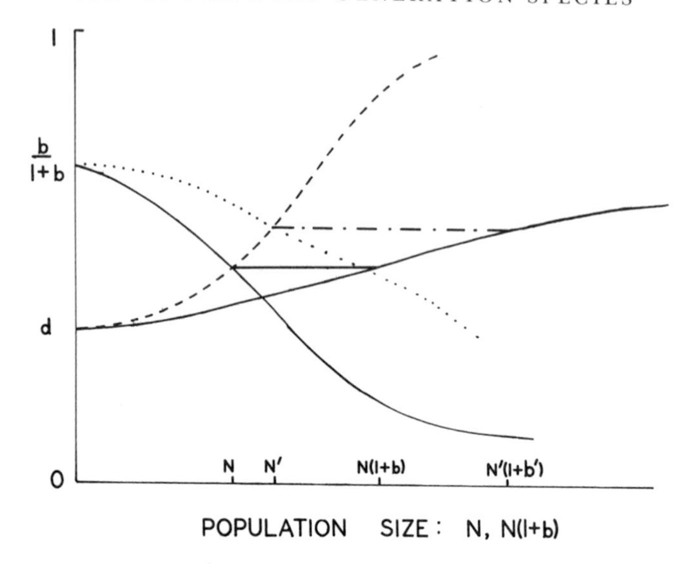

FIGURE 9. A change in breeding habitat abundance, when non-breeding survival is not sharply density-dependent.

there is a large shift. Since the curves are always either decreasing or increasing with increases in population, shifting them to the right has the effect respectively of raising or lowering them. Thus an increase in resource abundance in the breeding season raises the replacement rate curve except at $N = 0$, which is a density-independent maximum.

We can now consider graphically the immediate or short-term effects of changing resources in one season. First we define the phrase "limited by breeding (or wintering) resources" as follows: A population is at least partly limited by a season's resources if an increase in that season's habitat abundance effects an increase in the equilibrium of the population.

Figure 9 presents an example: the effects of an increase in breeding habitat. The dotted curve is the new g curve (see Figure 9), the solid curves the original g, d curves. Also given (dashed line) is the N transformation of the death-rate curve (denoted *d-N* curve). The ends of the horizontal dashed and solid line segments depict the new and old equilibrium populations. In Figure 9 the breeding population has increased from N to N' and the prewinter population from $N(1 + b)$ to $N'(1 + b')$. However, in Figure 10 another example is presented in which the *d-N* curve folds back. In this case, a given spring population N can obtain from three combinations of fall populations and associated winter death rates. Then, an *increase* in breeding habitat results in a *reduced* breeding population ($N' < N$). These figures are accurate and theoretically possible examples (see Figure 10). Thus, the relationship between habitat abundance and equilibrium breeding population size is not simple.

However, we can observe that raising the replacement-

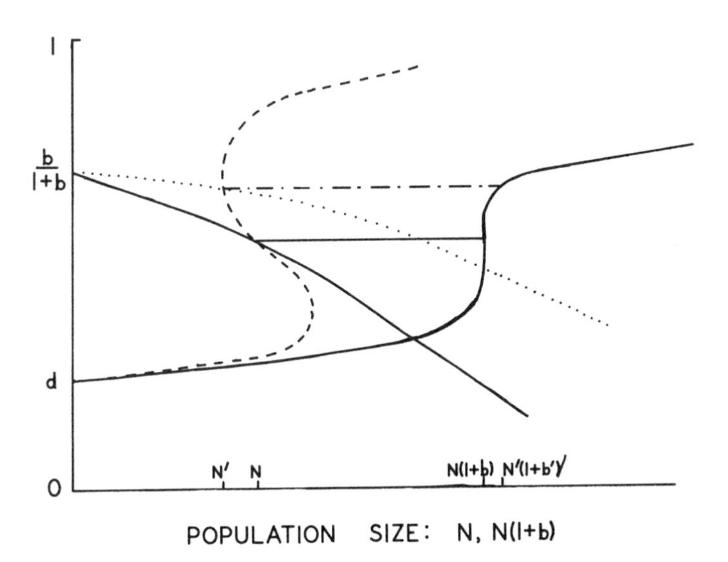

FIGURE 10. A change in habitat abundance during the breeding season when the *d-N* curve is folded back.

rate curve does raise the equilibrium replacement rate. So it follows that the death rate must also be raised. The death-rate curve remains unchanged, however, so a higher death rate implies a higher wintering population size. Thus $N(1+b)_{eq}$ increases when breeding habitat is increased. Therefore *winter population size is at least partly limited by breeding habitat.* A similar argument on N_{eq} when winter habitat is increased leads to the converse conclusion: *Breeding population size is at least partly limited by winter habitat.*

The degree of limitation, however, can be quite small. If at N_{eq} the replacement rate is near the maximal, density independent upper bound (g when $N = 0$), then an increase in breeding habitat cannot result in a large increase in replacement rate. Therefore, the new death rate cannot be considerably higher and the winter population not increased much. If we let the replacement rate *stay* at its

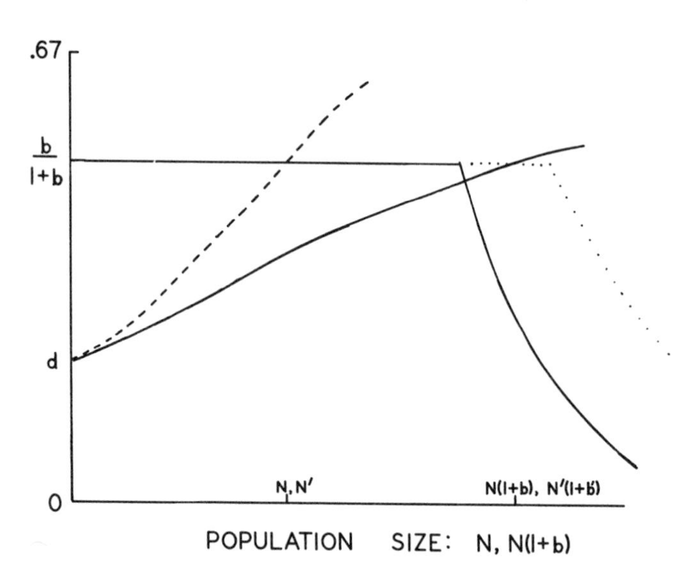

FIGURE 11. The case of a population limited strictly by winter resources.

maximum, we might find exception to the first generalization above. This is described in Figure 11, where neither the winter nor the breeding equilibrium populations is influenced by a change in abundance of breeding habitat. The intersection of d with the new g curve in Figure 11 is at the equilibrial value of $N(1 + b)$; neither g-$N(1 + b)$ transformation is shown, however, as these overlap the simple g curves in all regions of interest. This population is completely limited by winter habitat. A similar figure could depict a population completely limited by breeding habitat.

To summarize: Suppose (1) g and d curves are never flat, (2) g always decreases with the increase in N, and (3) d always increases with an increase in $N(1 + b)$. Then the breeding populations are partly limited by winter habitat, and winter populations are partly limited by breeding habitat. The degree of limitation by the opposing season's resources depends on how close the equilibrium rate of that season is to its low-density limit.

We have already noted that the effect of a season's resources on its own population size is complex. For the breeding population, of primary importance is the slope at equilibrium of the death-rate curve. It seems from Figures 9 and 10 that if the d curve is flat, the d-N transformation does not fold and the breeding resources will limit breeding population size. We prove this as follows.

First, we consider the case in which at equilibrium the replacement rate-transformation curve is not folded, permitting only one intersection. In this case a slight increase in breeding resources at population size N results in a slight increase in replacement rate, say to $g' = b'/(1 + b')$. There is, therefore, a slight increase in $N(1 + b)$ to $N(1 + b')$. We wish to know whether the new equilibrium breeding population, N', will be larger or smaller than the old value, N. Ordinarily, the pair $N, N(1 + b')$ will not be at equilibrium since the death rate at $N(1 + b')$ (denoted d') will not equal g'. Suppose it is higher ($d' > g'$). Then by lowering the breeding population from N, we shall raise the replacement rate from g'. Since the replacement rate-transformation curve is not folded, these new values of g and N will result in a lower prewinter population. This will lower the death rate. But, since we assumed $d' > g'$, gradually decreasing the death rate while increasing the replacement rate will make the difference between d' and g' smaller until it is zero and the two quantities are equal. Therefore, if $d' > g'$, the new equilibrium breeding population is smaller ($N' < N$). Conversely, if $d' < g'$, the new equilibrium breeding population is larger and breeding population may be limited by breeding resources. So, to examine the effect of breeding resources on breeding population, we must look more closely at d' and g'.

The relationship between d' and $g' = b'/(1 + b')$ depends on their mutual relationships to $N(1 + b)$, the fall population at the old equilibrium $d = g = b/(1 + b)$. The changes (d to d', g to g') are all reflected in the changes in $N(1 + b)$. We can compare the magnitudes of the changes in d and g by relating each of the variables to $N(1 + b)$. In particular, if, with a small change in $N(1 + b)$, $g = b/(1 + b)$ increases more than d, then $g' = b'/(1 + b')$ will be larger than d'. Conversely, if d increases more than $b/(1 + b)$, then $d' > b'/(1 + b)$. We therefore look at the derivatives of the two variables *re* $N(1 + b)$ (N fixed):

$$\frac{\mathrm{d}}{\mathrm{d}N(1+b)}\left[\frac{b}{1+b}\right] = \frac{1}{N}\frac{\mathrm{d}}{\mathrm{d}(1+b)}\left[1 - \frac{1}{1+b}\right]$$

$$= -\frac{1}{N}\frac{\mathrm{d}}{\mathrm{d}(1+b)}\left[\frac{1}{1+b}\right] = \frac{1}{N(1+b)^2}.$$

Thus, if

$$\frac{\mathrm{d}}{\mathrm{d}N(1+b)}[d] < \frac{1}{N(1+b)^2},$$

then $d' < g'$, and breeding populations may be limited by breeding resources. If

$$\frac{\mathrm{d}}{\mathrm{d}N(1+b)}[d] > \frac{1}{N(1+b)^2},$$

then breeding populations may not be limited by breeding resources. The value $1/[N(1+b)^2]$ thus sets an upper value on the slope of the d curve. If the d curve is too steep (slope $> 1/[N(1+b)^2]$) at the point of intersection, breeding resources cannot limit breeding populations.

This result, like that given in the Appendix to this chapter, determines the conditions under which the transformation curve is folded (see Figure 10). The present argument shows that if the rate curve of one season is folded while the other is not, and if the first season's curve is intersected just once in the region of the fold, then the resources of the first season do not limit the size of its own population.

The control of prewinter populations by wintering resources is analogous. If the birth rate is sharply declining, increases in winter resources per individual may lead to decreases in winter population.

It is clear from these arguments that the effects of (small) changes in resource abundance in a given season on the population of the same season are determined largely by the slope at equilibrium of the opposing season's transformation curve. Also important, however, is the population size at which one season's transformation curve reaches the density-independent (zero point) extreme of the other season's primary rate curve. Since a species will not improve on its density-independent extremes, excluding evolutionary changes, the population size at which one season's transformation curve equals the zero extreme of the other season represents a limiting population size. Increases in the latter season's resources can never push the population beyond this limiting population size. If the population is at this extreme equilibrium, further increases in resources can have no effects (Figure 11). This limit could be a maximum (nonfolded case), a local minimum (folded case, intersection in the region of the fold), or a local maximum (folded case, intersection above the fold).

SUMMARY

A species that has generation time about the same order of magnitude as a seasonal cycle, when exposed to a simple alternation of seasons, divides each generation into a season of birth and a season of death. By supposing that the births and deaths during a single generation can be simply related to two population sizes during the generation, hypothetical models can be generated which yield equilibrial population sizes for the species. If the dependence of birth and death rates on population size is too pronounced, however, several equilibria can exist at a single point in the life cycle, even with the same environmental seasonal cycle. Also, if the death rate rises too rapidly with density, small improvements in environmental factors affecting breeding actually decrease the equilibrial breeding population. The factors affecting a season's births or deaths are more certainly important in determining the next season's population and may have little effect on the season's own population.

APPENDIX

We can deduce what slopes of the replacement rate lead to transformation curves folded back. We are interested in $[dN(1+b)]/[d(b/(1+b))]$, the rate of change of the fall population, $N(1+b)$, with changes in g. If this derivative is positive, the transformation curve is folded back. Suppose the slope of g is α, so that

$$\frac{\mathrm{d}(b/(1+b))}{\mathrm{d}N} = \alpha,$$

then it can be shown that

$$\frac{\mathrm{d}b}{\mathrm{d}N} = \alpha(1+b)^2.$$

Also,

$$\frac{\mathrm{d}N(1+b)}{\mathrm{d}(b/(1+b))} = \frac{\mathrm{d}N(1+b)/\mathrm{d}N}{\mathrm{d}(b/(1+b))/\mathrm{d}N},$$

$$= \frac{1}{\alpha}\frac{\mathrm{d}N(1+b)}{\mathrm{d}N}$$

$$= \frac{1}{\alpha}\left[(1+b) + N\frac{\mathrm{d}b}{\mathrm{d}N}\right],$$

$$= \frac{1}{\alpha}\left[(1+b) + N\alpha(1+b)^2\right],$$

$$= \frac{(1+b)}{\alpha} + N(1+b)^2.$$

But

$$\frac{\mathrm{d}N(1+b)}{\mathrm{d}(b/(1+b))} = \frac{(1+b)}{\alpha} + N(1+b)^2 > 0$$

implies

$$N(1+b)^2 > -\frac{(1+b)}{\alpha}.$$

Since these terms are both positive ($\alpha < 0$), taking the inverses reverses the inequality

$$\frac{1}{N(1+b)^2} < -\frac{\alpha}{1+b}$$

or

$$\frac{1}{N(1+b)} < -\alpha.$$

In other terms, $-\alpha$ is the absolute value of the slope of the replacement-rate curve. If this value is greater than the inverse of the fall population, then the transformation curve is folding back, and several equilibrium populations are possible. Also (see text), if the d curve intersects the g-$N(1 + b)$ transformation curve just once in the region of the fold, then winter resources do not limit winter population.

For the death rate we are interested in

$$\frac{\mathrm{d}N}{\mathrm{d}d}.$$

If this derivative is negative, the transformation curve is folded back. Let

$$\frac{\mathrm{d}(d)}{\mathrm{d}N(1+b)} = B \qquad (B > 0).$$

Then

$$\frac{\mathrm{d}N}{\mathrm{d}d} = \frac{\mathrm{d}N/\mathrm{d}N(1+b)}{\mathrm{d}d/\mathrm{d}N(1+b)} = \frac{1}{B}\left[\frac{\mathrm{d}N}{\mathrm{d}N(1+b)}\right]$$

$$= \frac{1}{B}\,\frac{\mathrm{d}N(1+b)/(1+b)}{\mathrm{d}N(1+b)}$$

$$= \frac{1}{B}\left[\frac{1}{1+b} + N(1+b)\,\frac{\mathrm{d}(1/(1+b))}{\mathrm{d}N(1+b)}\right].$$

Since we are at equilibrium, $d = b/(1 + b)$, $1 - d = 1 - b/(1 + b) = 1/(1 + b)$, and $[\mathrm{d}(1/(1 + b))]/[\mathrm{d}N(1 + b)] = [\mathrm{d}(d)]/[\mathrm{d}N(1 + b)] = B$. Then

$$\frac{\mathrm{d}N}{\mathrm{d}d} = \frac{1}{B}\left[\frac{1}{1+b} - N(1+b)B\right]$$

$$= \left[\frac{1}{B}(N) - N^2(1+b)^2\right]\frac{1}{N(1+b)}.$$

If $\mathrm{d}N/\mathrm{d}d < 0$, then

$$\frac{N}{B} - N^2(1 + b)^2 < 0$$

or

$$B > \frac{1}{N(1 + b)^2}.$$

CHAPTER TWO

Theory for Organisms with Short-Generation Times

ASSUMPTIONS ABOUT THE SPECIES

The classical theory of populations is usually expressed in a continuous equation relating growth of population to population size, relative to some constant, K, which is thought of or defined as carrying capacity, for example, $\mathrm{d}N/\mathrm{d}t = r_0 N(1 - N/K)$. Seasonality can be introduced into such a system by writing K as a function of time that periodically rises and falls. I shall not attempt to describe such a system analytically, preferring instead to use a graphical approach.

Assume that K varies with time in some periodic way which creates an endlessly repeating cycle. Assume that in one of these cycles there is only one local maximum and one local minimum value. Consider $\mathrm{d}N/(N \cdot \mathrm{d}t) = \mathrm{d} \log N/\mathrm{d}t = \beta_0$, the net reproductive rate of the average individual. If N, the population size, is lower than K, then assume that $\mathrm{d}N/(N\mathrm{d}t) > 0$; if $N > K$, then $\mathrm{d}N/(N\mathrm{d}t) < 0$. Also assume that if $N < K$, $\mathrm{d}N/(N\mathrm{d}t)$ does not increase with a decrease in $K - N$, and that if $N > K$, $\mathrm{d}N/(N\mathrm{d}t)$ does not decrease with a decrease in $N - K$. Finally $N = K \Rightarrow \mathrm{d}N/(N \cdot \mathrm{d}t) = 0$.

These assumptions mean that if the population is below the carrying capacity, K, then N increases, but at a rate per organism that can only decrease (or stay the same) as the population approaches carrying capacity. If the population is above carrying capacity, it decreases but again at a rate per organism that in absolute value can only decrease as the population approaches carrying capacity. Individual reproductive rates for N may be temporarily density-independent, but there is some density-dependence in that these are always positive below K and negative above.

SOME APPARENTLY DENSITY-INDEPENDENT SYSTEMS

A simple system is shown in Figure 12 where N and K are plotted over time. During any one season, K is constant, abruptly changing at the end of one season to the next season's value. N can behave in several ways. It can simply move back and forth between K's (curve a of Figure 12) or it can be limited by K at one season, but never growing or declining enough to reach the K of the other season. For example, the population in curve b of Figure 12 is limited by summer K; the winter decline is so slow that the winter K is not even approached.

If population growth and decline are density-independent, the population may be controlled by the resources of one season. This depends on whether the

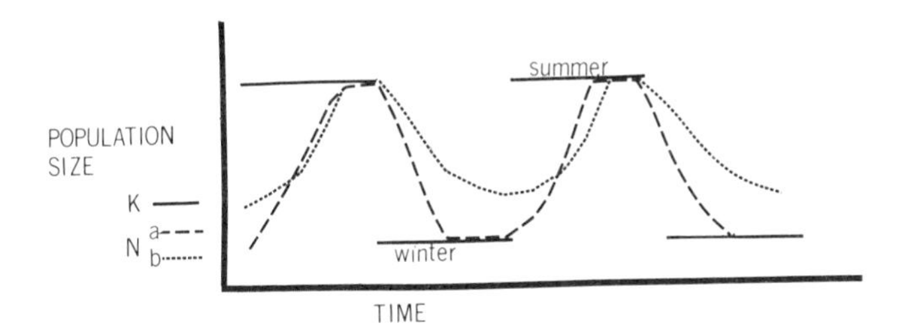

FIGURE 12. Population size as a function of time in a simple seasonal environment. In a, the population reaches both seasons' K's, simply bouncing between them. However, if the maximal decline of a population is less than the difference between K's, (b), the population will reach only the summer K, and so will be largely limited by summer resources.

maximal growth over the entire summer or the maximal death over the entire winter is greater or less than the differences between K's. In one case (curve b of Figure 12), the population stabilizes (in a deterministic model) by reaching the summer K each year, and staying there until winter. Then it suffers a certain winter loss which (1) is the maximum possible for the density-independent death rate and the time involved and (2) is less than the difference between K's. Growth in summer makes up the winter loss, but is less than the summer maximum growth, as K is again reached and growth stops before summer ends. At equilibrium, the summer growth is reduced to equal the winter attrition, and the population size at both seasons is regulated by the breeding K. Only changes in breeding resources can affect the population.

If the maximal growth over the entire summer is less than the maximal death during the winter and less than the difference between K's, the population size at both seasons is regulated by the winter K through analogous arguments.

Another simple system can be considered, again assuming primarily density-independent population changes (Figure 13). In this system, K continuously rises and then falls over a seasonal cycle. Below K, N rises such that $d \log N/dt$ is some constant; above K, N falls in a similarly constant way (constancy of $d \log N/dt$ defines density-independence). Changes in N must be low relative to changes in K or the population will merely follow the cycle of K.

In this system, what can vary is the relative length of the "breeding" to the "nonbreeding" season; at equilibrium, summer growth must again equal winter decline and the lengths of the seasons are varied to allow this. If $d \log N/dt = C_s$ in summer and $-C_w$ in winter, then the summer growth (expressed in logarithms) is $C_s t_s$ where t_s is the length of the summer growing season (as defined from

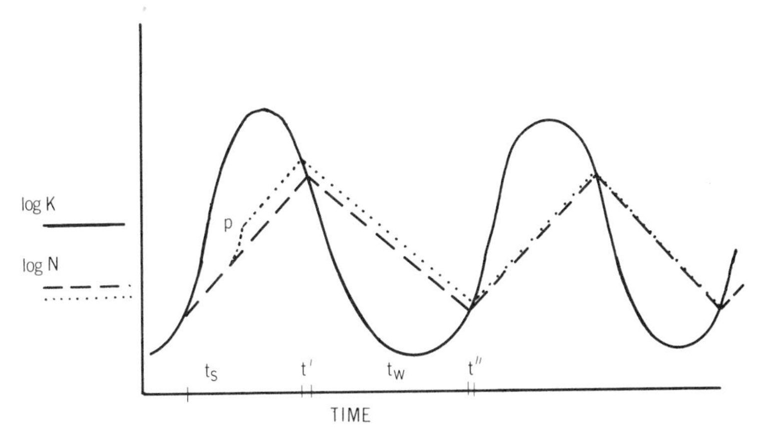

FIGURE 13. Population growth in a continuously fluctuating seasonal environment. The population shown is almost always under density-independent control, yet perturbations (P) always return to a stable equilibrium.

the time when the population starts growing until it stops). Winter decline is $C_w t_w$, t_w being the length of the winter. But $t_s + t_w$ must equal t_T, the length of the seasonal cycle which is constant. So we have two equilibrium equations:

$$t_s + t_w = t_T$$

and

$$C_s t_s - C_w t_w = 0,$$

which yield unique and specific values of t_s and t_w, as t_T, C_s, and C_w are given. This fixes the changes in log N with time, as noted in Figure 13, assuming of course that log N is a continuous function of time.

Perturbations in N converge back to this solution. For example, if in the summer there is a slight increase in N (P, Figure 13) then the population will encounter "winter" a bit sooner than usual (t_s is reduced by t') and will not grow so much. Should it still be above normal in

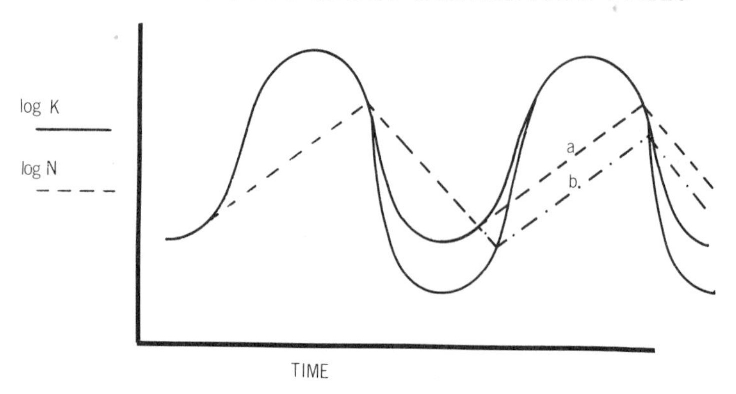

FIGURE 14. Population growth (density-independent) in a changing seasonal environment. The population cycle (– – –) is shown for the usual K cycle (a), and for a K cycle with lower winter K for the second and following winters (b).

the following winter, it will encounter spring a bit later (t_w is increased by t'') and will suffer more loss accordingly. And so the perturbed population will eventually reestablish equilibrium.

Such a population is almost always under density-independent control, but there will be density dependence if overall seasonal birth or death rates are computed using the theory of the previous chapter. And there are significant density-dependent effects during the spring or fall as N crosses K. A population will change with changes in the seasonal resources, depending on where it intercepts the K curve. Some examples are given in Figure 14, curves a and b.

DENSITY DEPENDENT SYSTEMS

More complicated, and far more realistic, is the system portrayed in Figure 15. Again K simply rises and then falls over a seasonal cycle, but periods when K is constant are permitted. Below K, N rises, but in this system, $d \log N/dt$

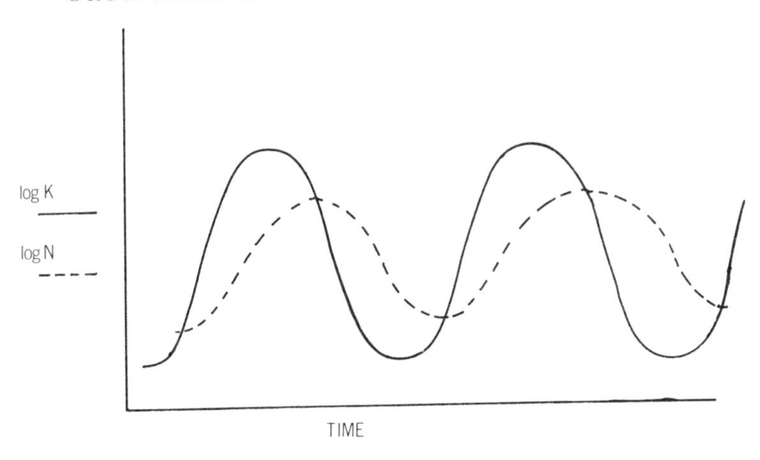

FIGURE 15. Density-dependent population growth in a seasonal environment.

is not constant but depends on $K - N$. Above K, N falls, again in a density dependent way. When $N = K$, $dN/dt = 0$. Since the rate of change of N varies, we shall be interested in situations where the growth and decline of N are small or large compared to changes in K per unit time.

In a seasonal environment, N might regularly deviate from K, due to seasonal shifts in the latter which occur at a rate that is faster than the population N can respond. The population may or may not "catch up" with K as a season progresses to maximum (or minimum) K. If it does not, then it must intercept K beyond the season's peak, as during this period K is changing in one direction and N in the other. For example, in the late spring N is greater than K, but K is increasing and N is decreasing, so the two quantities intersect at some point. (It is impossible in a deterministic model for N to have a value lower than $K_{min} = K_w$ or higher than $K_{max} = K_s$.) Beyond the intersection, K is still increasing perhaps faster than N can increase, and N again moves to "catch up."

This model is a rather easy extension of the previous one and leads to the obvious conclusion that whenever N catches K, the population is influenced by changes in the environmental factors that determine K at that point. If N is greatly removed from K, the changes in the environment which change K do not, relatively speaking, cause as great a change in $K - N$ and so the population is not as affected. Figure 16 describes populations limited by the K's of both seasons in combination.

We can observe directly that what is very important here is the rate of change of population relative to the rate of change of resources or resource determined K. In seasons when K increases very rapidly relative to N, the population may simply operate under the maximum population growth rate (r_0 in conventional terminology), suffering in this period no density dependent effects, including competition. Similarly, if K decreases very rapidly relative

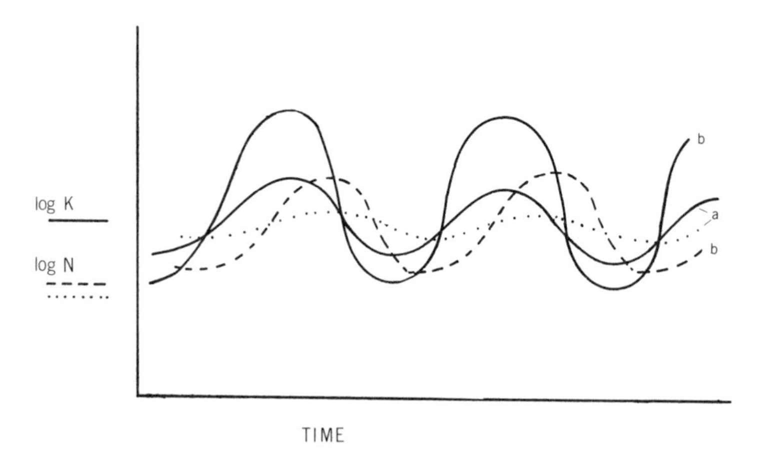

FIGURE 16. Density-dependent population growth in a seasonal environment. Both seasons' resources somewhat limit the population, in that changes in K change both the minimal and maximal population. But the population changes are not as extreme as the changes in K.

to N, the population may simply resort to some life form that has a fairly low, density independent mortality rate (e.g., resistant spores, eggs, or hibernating organisms).

Asides

I have raised the question: Can we *a priori* argue that a population will be equally limited by the two seasons, or limited primarily by the breeding season, or limited primarily by the winter season? Is one of the above alternatives more plausible than another? On the one hand, evolution will favor a species which maximizes its birth rate and reduces its death rate, forcing populations close to breeding K and keeping them far from winter K. However, it is generally easier to die than to reproduce. This would push populations to the *winter* K. Thus, we need a better understanding of the evolutionary rates of populations before this (in my opinion) interesting problem can be pursued further.

One wonders what in the present models corresponds to the S shaped transformation curves of the previous chapter. The answer is remarkably complicated. Results similar to those of the first part can be constructed here by having K depend not only on time but on N as well, particularly values of N at an earlier time (time lag effects). For example, K_{max} (late summer) could depend on N_{min} (N in the early spring); or K_{min} (spring) could depend on N_{max} (fall N). In the second case, suppose K_{min} decreased as N_{max} increased. That is, suppose a high fall population so overgrazed winter's resources that the damage extended into the following spring. Then K_{min} would be reduced by high N_{max} values. A population exposed to increasing K_{max} (more summer habitat) would increase N_{max} and so would decrease K_{min}. If it were winter limited and nearly "caught up" with the winter K_{min}, such an increase in breeding resources (K_{max}) would then lead to a decrease in

N_{min}, the initial breeding population! Such relationships may not be as likely in organisms with several generations per season, but should not be overlooked.

THE LIFE SYSTEM CONCEPT

Clark, Geier, Hughes, and Morris (1967) summarize a number of studies on various species of insects, primarily insect pests. Their method of summary utilizes a life system concept which in essence partitions the population trends into compartments according to the life cycle of the organism, much as I have partitioned according to seasonal variation in K. They, like I, have stressed the controlling influence of those particular periods, however defined, in which fitness is density-dependent. They have not formally developed their theory to the point where they can deduce from demographic curves equilibrium population levels. There is some merit in doing this here.

Suppose there are n stages of development in the life history denoted S_i, $i = 1, \ldots, n$. Each stage follows from a previous one, and we may suppose that the relationship between successive stages is density-dependent, so

$$\frac{N_i}{N_{i-1}} = f_{i-1}(N_{i-1})$$

where N_i is the size of the population of the ith stage. (Assume that if $i = 1$, $i - 1 = n$.) Thus, if $N_i/N_{i-1} > 1$, the process from stage i to $i + 1$ is reproduction; otherwise it is a period of decline or stability.

If we start at some stage of development (e.g., adults $= S_1$), we have some population present then, N_1. At the second stage of development, S_2, we have $N_2 = N_1 f_1(N_1)$. Also $N_3 = N_2 f_2(N_2)$, and so on, beyond the last stage of development where $N_{n+1} = N_1' = N_n f_n(N_n)$. If the population is stable, $N_1' = N_1$. Then

$$N_1' = N_n f_n(N_n) = N_{n-1} f_{n-1}(N_{n-1}) f_n(N_n) = \cdots$$

$$= N_1 \prod_{i=1}^{n} f_i(N_i), \tag{1}$$

so

$$1 = \prod_{i=1}^{n} f_i(N_i).$$

However, N_i depends on N_{i-1}, which depends on N_{i-2}, which . . . and so on, which depends on N_1, so each term in the above product can be written as some function of N_1. For example, $N_2 = N_1 f_1(N_1)$, so $f_2(N_2) = f_2(N_1 f_1(N_1))$. Also $N_3 = N_2 f_2(N_2) = (N_1 f_1(N_1))(f_2(N_1 f_1(N_1)))$. If all the terms of the product (1) are written in terms of N_1, then (1) is an equation in N_1 which can be solved for N_1. There may be no or many solutions for this equation, depending on the component density-dependent relation-

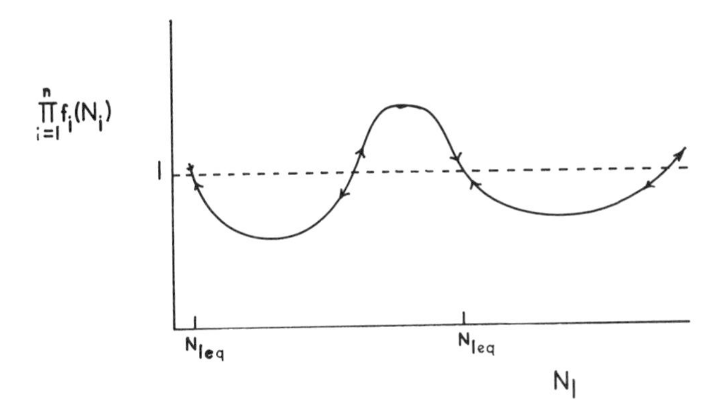

FIGURE 17. The product, $\prod_{i=1}^{n} f_i(N_i)$ yields the relative population growth over a life cycle involving several stages. At 1, this product indicates equilibrium. It may be a very complicated function, however, with several equilibrium points, if there are many life stages, all with density-dependent birth or death. Those equilibria with negative derivatives (converging arrows) are stable.

ships (the $f_i(N_i)$). In the simplest case, if one of the $f_i(N_i)$ is linear in N_i, and the rest are constant functions, (1) can be solved uniquely in N_i. If all the $f_i(N_i)$ are linear in N_i, then there will be $\Sigma 2^i$ solutions for N. Some (about half) of these solutions are unstable. This may be determined by taking the derivative of (1) and examining the sign of this derivative at each of the values of N obtained from (1) directly. Only equilibrium values of N with negative derivatives are stable (see Figure 17). Also, some solutions of (1) may be negative and can also be ignored.

It is apparent from this development that an organism (Figure 17) with a sequence of life stages that are dependent on one another mainly through changes in population size might have several stable equilibria in population size for each life stage, and might show variation in population size through changes in the equilibrium size. Such variation need depend only on simple basic relationships.

SUMMARY

When a species has several generations during a single season, there is a closer fit to the classical continuous logistic equation describing population growth. By letting K in this equation be a function of time, the logistic and similar models can describe a population living in a seasonal cycle. A graphical analysis of such models yields systems in which there is almost no density dependence yet which show unique and stable population size fluctuations. Systems in which density dependence is found also have simple, unique, stable population size fluctuations. The sizes of these populations can be largely determined by the values of K during just part of the seasonal cycle, depending on how rapidly K increases or decreases with time, relative to the rate at which N can similarly change. A population near K with a low r_0, confronted with a rapidly increasing K, may after a short time find itself so far from K that further changes in K do not affect it.

A model for populations with several sequential life stages is developed indicating that nonunique equilibria may exist in these cases.

CHAPTER THREE

Applications of Graphical Models

GENERAL

The major application of these models is toward the understanding of control of specific populations. Examples are given below. Each model is one statement (there are others) of the way that numbers are adjusted to the environment. They have the virtue of separating the effects of the different seasons so that changes in resource abundance at different times of year (e.g., the effect of winter feeding stations on bird populations) can be evaluated. They are also expressed in the form of curves that are well defined and perhaps not too difficult to measure.

The theory applies to studies other than population control as well. For example, consider the study of competition. It is clear that there is no *a priori* reason why either population growth during the breeding season or nonbreeding season population decline should be very density-dependent for the population as a whole. Although both of these rates may be density-dependent, only one need be for stable equilibrium. Thus, there is no justification (besides convenience) for assuming without evidence that, for instance, breeding space is limited. This point is important to studies of competition, since a lack of density-dependence in a given season implies a lack of meaningful intraspecific competition during that season, and therefore, of interspecific competition as well. The significance of a study of competition in the breeding season when it is winter food that is limiting (and vice versa) is therefore open to question. Most studies of competition in birds, for example, are conducted on breeding populations. However, Lack (1966), after reviewing bird population research suggests that it is more probable that winter (nonbreeding) resources limit most bird populations. Also, there exists seasonal variation in species overlap of food resources (Newton, 1967; Zaret and Rand, 1971) so that during certain seasons competing species use almost completely different resources, while at other seasons they overlap considerably. This again implies that just one season may be limiting populations and that competition should be studied then.

The theory should also be important to students of evolution (e.g., adaptive morphology). Species limited largely by one season might be expected to evolve differently from those limited by the other. Selander (1965) suggests, for example, that sexual polymorphism in body size in birds is related to competition for breeding territories, which is more likely to occur if a species is limited by breeding space. Also, if a species varies its food or feeding behavior with the seasons, one might expect the feeding adaptations to be more closely adapted to the more limiting season. For example, Grant (1966) has developed the idea that the tarsus length of birds is positively adapted to the stability of the perches used while feeding. The winter season data presented by Grant agree with his principle, but Engels (1940), Osterhaus (1962), and Grant (1966) present data obtained from studies in the breeding season which do not agree with the principle.

In summary, resources vary with the season. The pattern of variation can be different in ways that have far-reaching consequences to the biology of the organisms using those resources. Studies of population size are particularly concerned with these relationships, but studies of competition and evolution may also be critically affected. The present models demonstrate ways in which season and resources can interact to affect populations of certain species.

SPECIFIC APPLICATIONS

Data from a number of population studies will be analyzed in the way suggested by the models. I have attempted to consider a wide variety of organism types.

The Great Tit in Marley Wood

Lack (1966) summarizes data for population changes of the great tit (*Parus major*) in Marley Wood near Oxford, England. The model offered above provides a way of looking at those data in order to answer the question of how tit populations are controlled. Two precautionary remarks must be made, however. The first is that Lack's data are from a single habitat, and the prime habitat of the species at that. Population changes are sometimes best reflected in secondary habitat types where variation is more pronounced (Kluyver and Tinbergen, 1953). Also, dispersion out of the habitat must be treated as mortality, and this is not necessarily the proper way of handling this phenomenon.

The data are presented in Table 1 and Figure 18. Generation time in the great tit is one year, the same order of magnitude as the annual seasonal cycle, so the first models are appropriate (Figure 7). Solid circles represent replacement rates, and solid squares and stars represent death rates. Small squares and stars are death rates in low beechmast years, medium squares and stars in medium beechmast years, and large symbols in high beechmast years. The death rates are divided into two groups, the first representing data from 1947 to 1955 (squares), the second representing data from 1956 to 1963 (stars).

TABLE 1. The great tit in Marley Wood

N	b	$b/(1+b)$	$N(1+b)$	d	Beechmast rank
14	4.60	.82	79	.47	2
42	4.95	.83	251	.76	8
60	4.35	.81	320	.81	1
62	3.05	.75	252	.75	8
64	2.45	.71	221	.82	0
40	3.75	.79	190	.78	3
42	4.40	.81	226	.73	5
62	4.25	.81	324	.83	0
54	3.40	.77	237	.80	0
48	3.95	.80	237	.59	6
98	2.30	.70	323	.83	0
54	2.80	.73	204	.60	4
82	3.40	.77	359	.72	0
102	2.70	.73	376	.53	5
172	2.10	.67	537	.83	0
86	2.65	.72	312	.75	4
78	3.25	.76	331	.67	0
110	–	–	–	–	–

From Lack (1966).

Three mortality, prewinter population size curves are drawn, one for high and low beechmast crops in the first group of years, and the curve for low beechmast crops in the first group of years, and the curve for low beechmast in the latter group of years. The separation by beechnut crop is justified by the fact that the species utilizes this food in the winter. The separation by years appears in the data (stars to the right of squares), and must be due to changes in the environment (e.g., the building nearby of a housing development, Lack, 1966), or to a change in study techniques (Perrins, pers. comm.). Only one replacement curve is drawn. Also, the transformation curve of the d curve from the low beechmast, pre-1955 data is drawn. This d transformation nearly folds back, indicating that breeding resources cannot limit breeding population. The g transformation is only slightly density-dependent, so

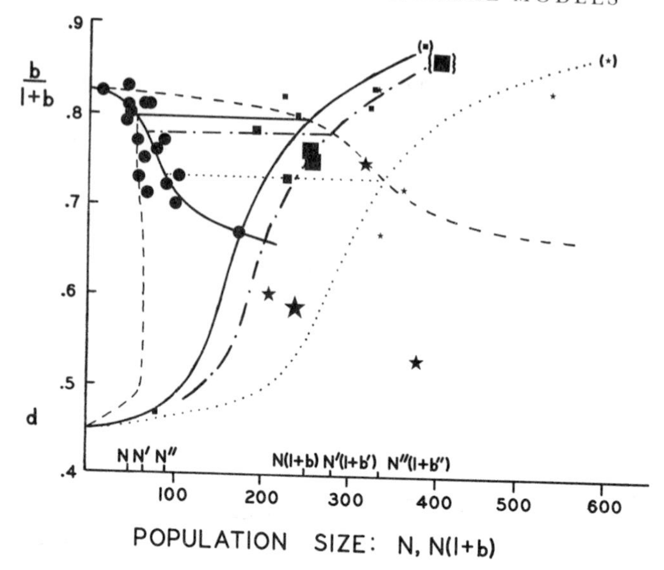

FIGURE 18. The great tit in Marley Wood: a scatter plot of Table 1. The curves are hypothetical, based on arguments in the text. The round dots are birth-rate (g) values, the squares and stars are death rates. Death rates are grouped by period (squares are early years of study, stars are late years), and by beechmast (small symbols mean low beechmast, large figures mean high beechmast). The bracketed symbols identify hypothetical curves associated with a certain period and beechmast crop. Thus, the fact that the small-star curve is displaced to the right of the small-square curve means that more habitat or resource was available later on in the study. The vertical dashed line to the left is the d versus N transformation, for low beechmast in the early years' regulation.

breeding resources on the average affect winter population sizes only slightly. Of course, during high beechmast years and later on in the study, the apparent increase in winter resources would yield different curves with different implications.

From the preceding theory and Figure 18 it seems that the great tit in Marley Wood is largely limited by winter resources. The equilibrium replacement rates are density-dependent, but are also much nearer the density-independent extreme, so changes in breeding resources would effect relatively little change in the prewintering or breeding population. But, as seen from the curves, changing winter resources (e.g., increasing beechmast) markedly change the breeding and prewintering populations.

The conclusion from the analysis as presented in Figure 18 is in essential agreement with the intuitive feelings of those who have worked with this species (Lack, 1966, p. 68), and may be regarded as little more than a formal presentation of that intuition. Lack (op. cit., p. 76) has regarded the density-dependence of winter mortality as essentially nonexistent however, and therefore considered the problem of population regulation in the great tit unsolved by the above data. The points plotted in Figure 18, however, indicate that density-dependence in nonbreeding mortality may well exist when the variations in beechmast and environment are taken into account. The lowest prewinter density has the lowest death rate and the highest prewinter density, one of the highest death rates. This in itself is rather unlikely to occur by chance with so many points (about one chance in a hundred). The scatter is pronounced, however, and no conclusion is certain.

I should observe that the death-rate curves are hypothetical, being my hypothesis of the particular system that generated the data scatter. There are some supportive arguments for this particular hypothesis which will be presented later. However, note that the broad and diffuse scatter is about what would be expected from a set of sigmoid-shaped curves that are horizontally displaced.

Figure 19 presents an alternative analysis of the great tit data, leading to a quite different interpretation. In particular, in some years Lack provides estimates of the winter population in December, and it can be argued that this is $N(1 + b)$, not the maximum summer population as used above in Figure 18. In Figure 19, the population as a whole

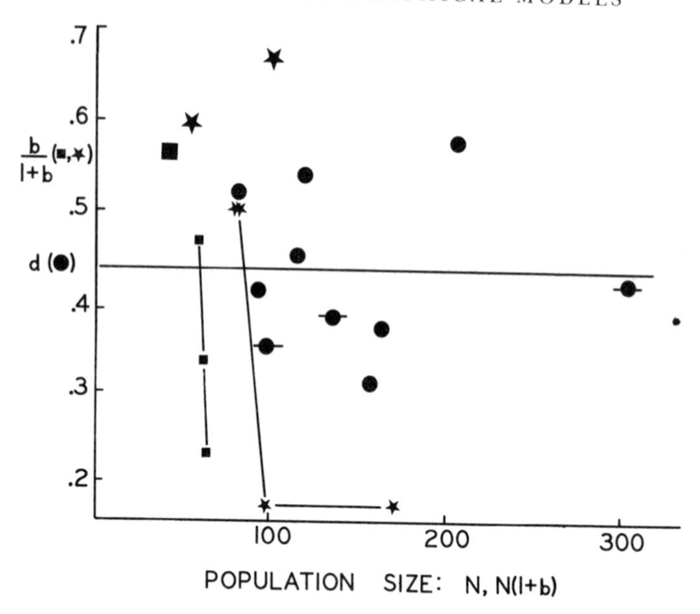

FIGURE 19. Alternative analysis of great tit data, where fall or late summer mortality is considered part of the breeding season, stars and squares are birth rates, stars reflecting higher beechmast years. Circles represent death rates, December through April; those with lines are high beechmast years.

appears to be limited by breeding (including fall) and not winter resources. The density-dependence regulating the population is in this case largely determined by some factors affecting breeding success plus some factor correlated with the beechmast crop that fall. Lack maintains that the bulk of this regulation occurs shortly after fledging, before the beechmast falls. So, it is not obvious why the beechmast is a significant correlate. Beechmast crops are cyclic, varying independently of the environment (Lack, 1966, p. 66).

The first (Figure 18) and I think more probable analysis, if in fact true, may be of considerable practical importance. It is generally conceded (Lack, 1966) that breeding birds (including tits) do not greatly affect the abundance of their insect prey and are thus useless for the control of insect "pests." The analysis presented above suggests that this is only because the birds are so limited by winter food that breeding densities are held below an effective insect-controlling level. This difficulty could therefore be relieved by providing consistent, extensive sources of winter food from late summer to spring. Assuming that this food would be used (a safe assumption, for most tit species at least) and that nesting holes are not limited (i.e., nest boxes would be supplied), the breeding density of tits could be increased in woodland areas to any level needed for insect control. This suggestion, although not particularly novel, is easily testable.

The sigmoid nature of the d, $N(1 + b)$ curves in this figure is not completely justified by the data, which are extremely variable. However, Figure 19 shows that at winter (December) population sizes ranging up to 200, d is not noticeably density-dependent, and has an average value of about .45. Therefore, the curves in Figure 18 are all held at about this value of d, for $N(1 + b)$ up to around 100 to 200. It seems from Figure 18, however, that for $N(1 + b) > 250$, d is, with some exceptions, much higher but is again not noticeably density-dependent. Thus, sigmoid curves are drawn to fit the d, $N(1 + b)$ data points, the sigmoid curves being flat at both high and low values of $N(1 + b)$, but rising steeply at intermediate values.

The Studies of Errington

Errington (1945) provided a remarkable summary of population regulation in the bob-white and later (1957) in the muskrat. His data are quite appropriate to the present models for long-generation species and so are here analyzed to provide further examples.

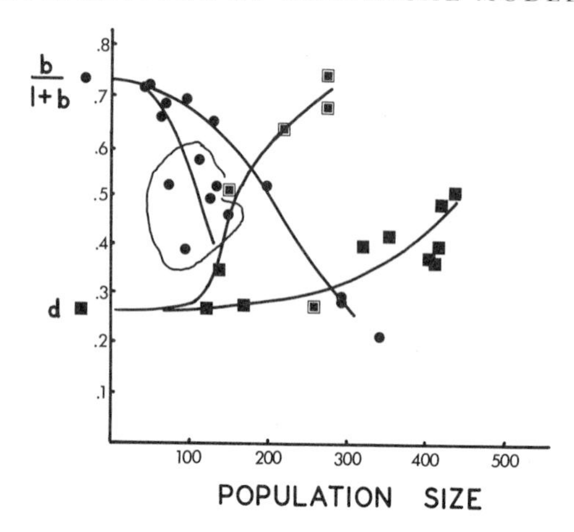

FIGURE 20. Errington's bob-white data in the long-generation analysis. I have drawn in two hypothetical sets of curves that might have generated some of the data points.

The bob-white quail in Prairie du Sac. Errington separated the winter mortality in his study into emergency (density-independent?) and nonemergency (density-dependent) losses. This helps reduce the variance in our analysis. I have computed the average density-independent loss (20.7 percent), and added to this the density-dependent effects. The results are plotted (Figure 20). Following Errington, I have divided the winters into two types, separated largely on the basis of where the points fall, and have drawn two *d* curves. Similarly, the breeding production data appear to divide (roughly) into two groups, from different kinds of summers. Again, appropriate curves have been drawn by eye. These curves were then lifted onto new graphs (Figure 21) and the transformation curves were drawn.

Only two of the four possible combinations are presented; the other two combinations of kinds of season evidently do not occur together. For example, five of the six below-average *g* points (Figure 20, enclosed by an irregular line) are from the summers preceding the winters of the above-average d points. The exceptional point is the least below-average and could easily be a random deviate from the high group norm. There are eight above-average d points; five of these, as noted (enclosed squares) were preceded by a summer with a low g; the remaining three were preceded by summers with such low density that the summer g curves cannot be separated. Thus, there is no good evidence contradicting the apparent relationship between low g summers followed by high d winters. The linear correlation between g and following d is $-.505$ (17 d.f., $p < .025$) (g is corrected for N, so that at a given value of the spring population a low birth rate is correlated with a high death rate in the following winter). The birth rate is not well correlated with the death rate of the preceding winter ($r = .25$, 17 d.f.). If, in fact, these correlations between winter survival and preceding breeding success hold, then

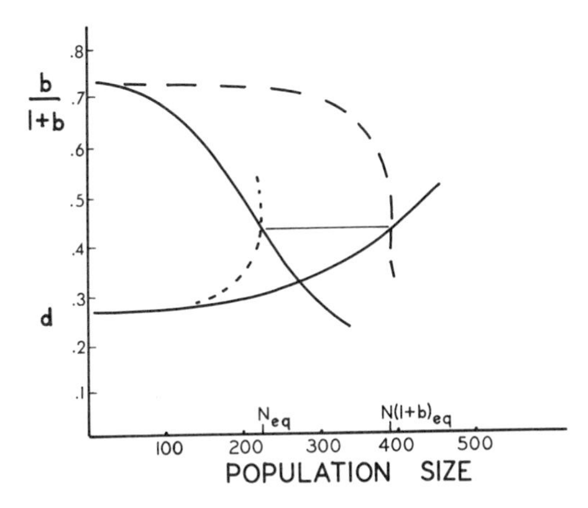

FIGURE 21. This figure designates the hypothetical curves and their transformations for mild winters. The equilibria are shown, but the actual population fluctuated so drastically that the equilibria were usually not even approximated. However, the values do provide interesting expected values.

there might be a breakdown of the theory. Either d or g might depend on the other or on factors primarily affecting the other. For example, if when g is high, the fall population is in better condition than when g is low, the winter survival is appreciably higher (d lower). Then two winters could exist which were identical in all respects, including population size, yet with different expected rates of death. This is not considered under the present theory. Or alternatively, if the population anticipated a high d in the forthcoming winter and there was a curtailment of reproduction (Wynne-Edwards, 1962), g would depend on factors affecting d, and identical summers could have different expected growth rates. In such cases, the simple plots of Figures 20 and 21 cannot tell us which season(s) is (are) limiting.

Of course, it is also possible that environmental factors producing a low g carry over negatively to the winter so the winter has a high d. In this particular study, Errington measured reproductive success as of November 1. It is possible that the regulating mortality in quail, as in the great tit, occurs over the fall, August to December. If winter food is limited, a shortage could easily be effective (through behavioral interactions) in October when fall food production is nearly complete, so dividing the growth and death seasons in November would assign half this fall mortality to summer g and half to winter d. Reproductive success would then be correlated with winter death.

There are, of course, other explanations for these curves. The above discussion is designed to emphasize the shortcomings of the present theory as well as those of certain kinds of data.

In concluding consideration of Errington's bob-white data, note that I have said very little that Errington himself did not say, or at least imply. In this particular case, my analysis little more than summarizes, defines, and places in a wider context the original interpretation of the data.

The muskrat in north-central Iowa. Errington's muskrat data are similar in nature to the quail data, consisting largely of November 1 and April 1 censuses. There are two additional features, however. Errington divides the data into two habitats, which can therefore be analyzed separately. Also, there are data on factors affecting nesting success which can be used to assess more carefully the density-dependent factors limiting the population.

The data are plotted in Figures 22, 23, and 24; one for stream habitats, one for marsh habitats, and one for combined population. The stream habitat population appears to be strictly winter limited. However, the marsh habitat population demonstrates significant density dependence primarily in the breeding season while the combined population appears limited by both seasons.

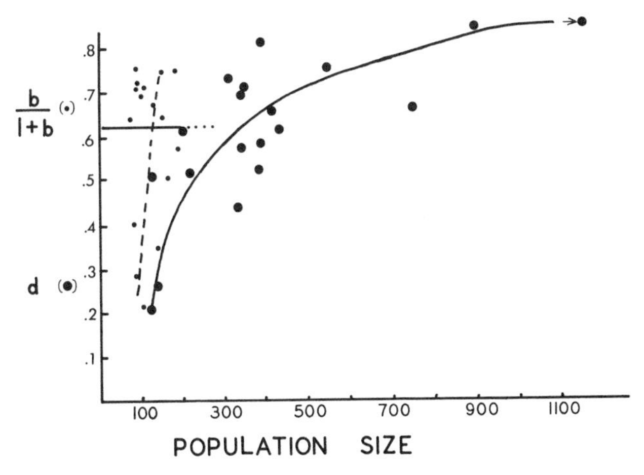

FIGURE 22. Stream muskrats. Small dots are $g = b/(1 + b)$ versus N, large dots are d versus $N(1 + b)$. Dashed line is the d-N transformation. The population appears totally limited by nonbreeding resources because the d-N transformation is essentially vertical. Variation is extremely great, however.

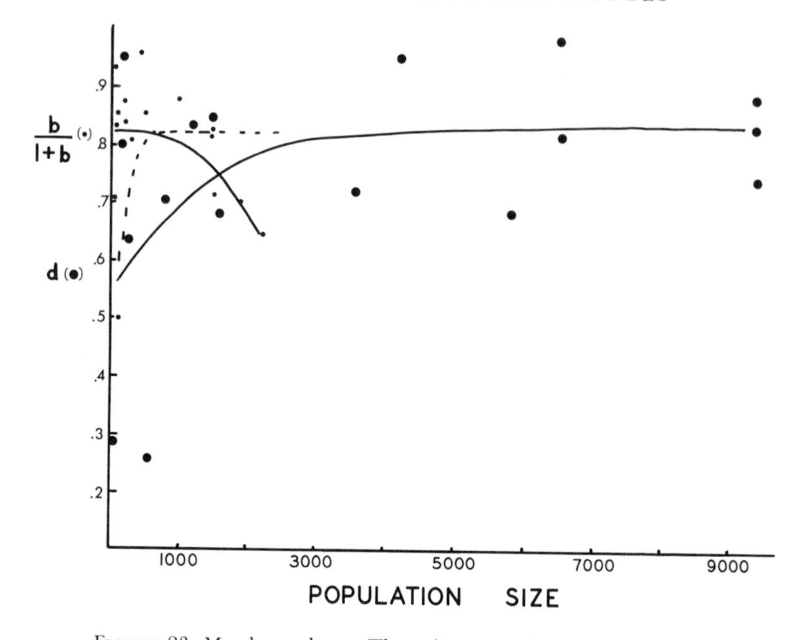

FIGURE 23. Marsh muskrats. There is some evidence of density-dependent breeding success and the death rate asymptotes at a value near the g high value. Very little change is needed to make the population strictly breeding-limited.

As in the bob-white, shifts in the winter curve for the combined population are correlated with shifts in the breeding curve, indicating again some breakdown of the theory or an inappropriate division of the seasons. When g and d are corrected for N and $N(1 + b)$, g is better correlated with the succeeding winter's d ($r = -.34$, 14 d.f.) than with the preceding winter's d ($r = .10$, 14 d.f.). Again, it seems that either during the breeding season conditioning affects winter survival or factors affecting the birth rate also affect the death rate.

In the original work (Errington, 1957) data on litter size are given. However, the relative changes in litter size are very small compared to the relative changes in birth rate.

This indicates that the important factors regulating birth rate are not all operative during the early part of the breeding season but that some operate at the end. With this evidence it seems advisable to advocate the hypothesis that fall mortality is important in regulating the population, affecting measured *g* and *d* alike. Then there would be little density-dependence of reproductive success (due to non-breeding, litter success, or litter size) up to late August; the apparent density-dependence in Figure 23, by this hypothesis, would be due to largely density-dependent fall mortality. This would explain the correlation between *g* and *d*. It would have been better, perhaps, to conduct censuses in September as well.

The data for the two habitats are intriguing in that they

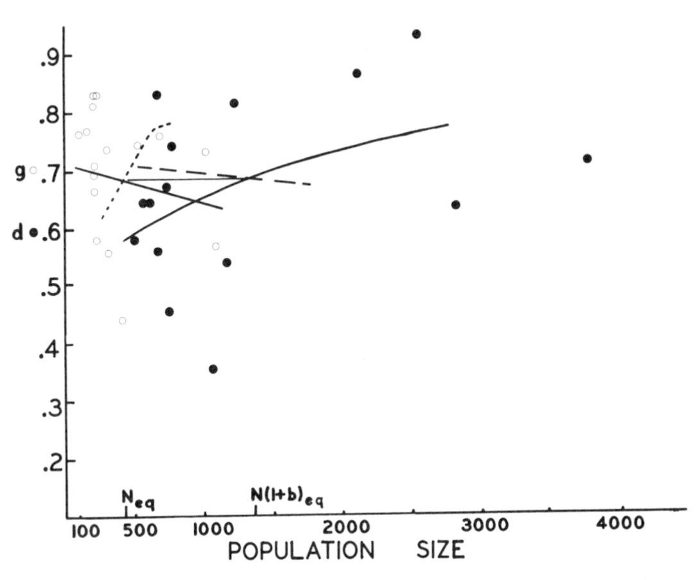

FIGURE 24. Stream and marsh muskrats. The finely dashed line is the *d*-*N* transformation, the broadly dashed line the *g*-$N(1 + b)$ transformation. The winter seems somewhat more important than the breeding season, but evidently both are limiting.

indicate the presence of different regulatory mechanisms for separate populations within a single species. There are unanswered questions about the extent and nature of interaction of these two populations, or conversely, the extent of their independence. It seems to me that the description and study of within-species variation in demographic parameters is currently a field of research which is very relevant to our understanding of human populations. This phenomenon has also been observed in insects (Clarke et al., 1967).

The Population Trends of Thrips imaginis

As an example of an application of the present theory to a short generation organism, we shall look at the population ecology of *Thrips imaginis* as described by Davidson and Andrewartha in 1947a, b.

Note that the long-generation theory and analysis concern only peaks of population growth, and so consider only populations at the beginnings and ends of seasons. The short-generation theory, on the other hand, concerns population growth *during* a season and is not especially concerned with peaks.

Thrips are small insects which live in roses and weeds, in the flowers. The eggs are laid in flower tissues and the larvae grow up there. Pupation is in the ground litter, however. Davidson and Andrewartha measured the average number of *Thrips* per rose nearly every day for six years. They have data from early spring for 14 years.

In the section on regulation of populations with short generation times, much emphasis was placed on logarithmic plots of N versus time. Curvilinearity in such plots indicated a nonconstant growth rate which could be related to density-dependence. So we would like a plot of log numbers of *Thrips* versus time. Davidson and Andrewartha offer monthly averages for the initial six years and log averages for the total 14 years, spring data only. Conveniently, they also provide polynomial regressions of log N versus time: t, t^2, and t^3.

Log N depends significantly on the higher order terms during the period of maximal spring growth, indicating curvilinearity, so density-dependence is at least possible. We would expect from a density-dependent argument that the curvature of log N with time be convex; the rate of growth should decline, on the average, as the population grows and approaches K. This is indicated in the average spring growth equation:

$$\log Y = 1.6948 + .038t - .00042t^2 - .00009t^3.$$

The higher terms in t are both negative, establishing convexity, supporting the hypothesis of density-dependence. Smith (1961) has also found density-dependence during this period.

The presence of density-dependence in the period of spring growth, where it is least likely and where the population is farthest from K, allows us to proceed assuming a logistic-type differential model. However, it does not disprove Davidson and Andrewartha's claim that density-dependence is not important in regulating the population during this season. Recall that the tits in Marley Wood demonstrated density-dependence most clearly in the breeding season but were more likely limited by the winter season. The question is not how density-dependent the population is but how close the population is to K. The population close to K is apt to fluctuate as K fluctuates, showing a relatively small *density* component in its variation. Only the population which is far from K and therefore little influenced by K fluctuations is apt to be free enough from these environmental fluctuations to show *density* correlations. Thus, Smith's (1961) clear demonstration of spring density-dependence in the present *Thrips* data perhaps only

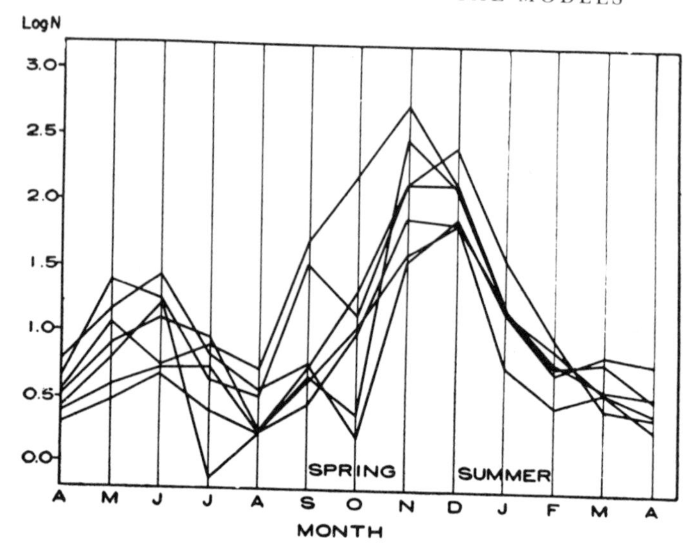

FIGURE 25. The logarithm of the average population size per month for 81 consecutive months in the study of *Thrips imaginis*. (Smith, 1962, drew this figure from the data of Davidson and Andrewartha.)

supports Davidson and Andrewartha's claim that N was far away from K, and therefore relatively free from density-dependent control.

To provide an overview of the *Thrips* data, log N versus time plots are given in Figure 25. We do not know what K is, but we may surmise, as did Davidson and Andrewartha, that the population is near K during most of the late summer and winter (January to September). This is generally a period of decline and, except for midsummer, the rates of change of log N are moderate. High potential rates of growth and decline are also indicated (spring and midsummer respectively), indicating that during these periods the population is farther from K. An educated guess, no better than that of Davidson and Andrewartha (1947b, p. 221), is that the winter K limits the population. The spring K

probably goes out of reach of N and the population then grows at a near maximal rate until the late summer drought abruptly ends the period of growth, dropping K to near its winter value. The population then declines. The peak of spring growth depends on the winter K and on the length of the growing season. When the growing season is very long or when the winter K is very high, the spring population perhaps gets near enough to the spring K to show indications of reduced growth rate.

The reader is, I trust, dissatisfied with this overview. There is a compelling need to know K. I shall offer two approaches toward filling this need. The first is quite crude but supports the view given above.

Population growth can be described with the logistic differential

$$N' = \frac{dN}{dt} = r_0 N \left(\frac{K - N}{K} \right).$$

We shall use this equation to approximate reality:

$$\frac{N'}{N} = \frac{d \log N}{dt} = r_0 \left(\frac{K - N}{K} \right) = r_0 - \frac{r_0}{K} (N).$$

Approximating $(\log N)'$ with $\Delta \log N$, we find ourselves interested in regressions where $\Delta \log N$ is correlated with N. Clearly, the intercept of this regression estimates r_0 and the slope divided into the intercept estimates (slightly biased) K. The study of density-dependence via these sorts of correlation has been challenged by Eberhardt (1970). Eberhardt argued that even if N_{t+1} is density (N_t) independent, $(N_{t+1} - N_t)/N_t$ will be negatively correlated with N_t. The conclusion is correct, but there is some misunderstanding of the meaning of density-dependence. Density-dependence is a phenomenon *not* of population sizes, but of population growth rates. In fact, the assumption that N_{t+1} is independent of N_t is itself a density-dependent assumption.

N_{t+1} is derived from N_t by population growth, and these entities cannot be independent. They may appear so only if r, the growth rate connecting them, is negatively dependent on N_t, i.e., is density-dependent. If, in fact, r is independent of N_t, then clearly the correlation of $(N_{t+1} - N_t)/N_t = [(r - 1)N_t]/N_t = r - 1$, and N_t will be zero, and N_{t+1} will be positively correlated with N_t. If N_{t+1} is independent of N_t, the population size is nicely tuned to current environmental conditions, and excesses or deficiencies in the previous time period, t, have been corrected for by adjustments in r.

Before considering this regression we must decide what time period to consider. Davidson and Andrewartha liked to consider time by counting backwards from the peak of spring population growth. They felt that peaks at different dates represented different kinds of seasons which could be aligned using the *Thrips* population as a phenology indicator. However, as seen in curve A of Figure 26a, this interpretation is not always correct. Clearly populations which get different starts in the early spring, and which

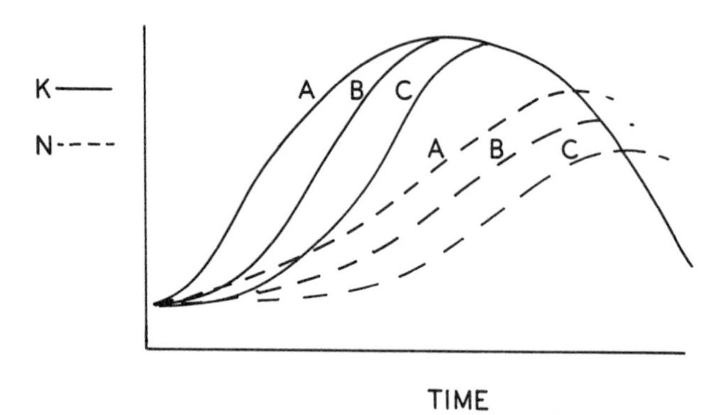

FIGURE 26a. Short-generation population growth when spring rise in K is variable but fall decline is stable. A, B, and C are in order of earliness of spring. Note that the earlier the spring, the earlier the population peaks, and the larger the maximal value.

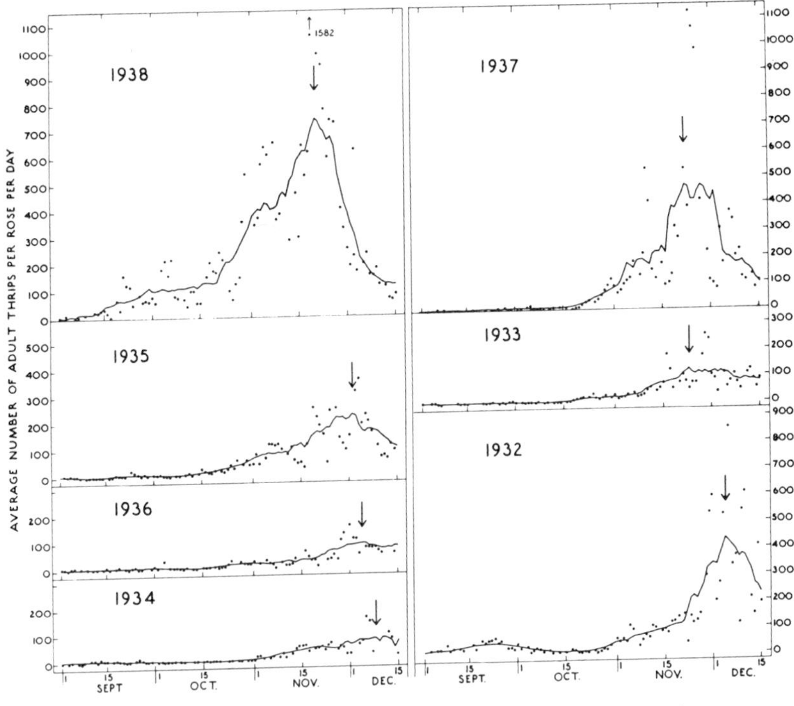

FIGURE 26b. *Thrips* population growth when spring is more variable than fall. Given is the number of *Thrips imaginis* per rose during the spring, each year for seven consecutive years. The points represent daily records; the curve is a 15-point moving average. The arrow indicates the data on which the curve reaches the maximum for each year. Andrewartha and Birch (1954, p. 576) state, "The beginning (of spring) is more variable than its ending. . . ." Note that lower maxima are achieved later, as predicted by Figure 26. (After Davidson and Andrewartha, 1947a, b.)

therefore peak at variable heights, will also peak at variable times if K falls at all gradually, as indeed it must. The test of this hypothesis is a negative correlation between peak height and peak date, which is amply verified by Figure 26b from the original report (Davidson and Andrewartha, 1947b, p. 196). The correlation of peak height and date is $-.82$ and the regression slope is -31 *Thrips*/day (p. 196).

This slope, by Figure 26, is the slope of K decreasing with the decline of summer. Thus, the use of peak dates to co-ordinate seasons is perhaps not justified and we shall not use this approach. We shall rely only on simple dates.

In actually making the correlations, I have computed the average N between two months and ΔN. I have approximated $\Delta \log N$ with $\Delta N/N$, and N with $\bar{N}$. This is very rough but the present analysis is designed to be more suggestive than conclusive.

Table 2 gives some estimates of r_0 and K for Davidson and Andrewartha's data, as presented in Table 3, p. 197, in their original work. All pairs of the months are not presented as the correlations between Y and X in these cases were either not large or not negative (in either case, presumably N is near K). Replication is over years; thus there are only five degrees of freedom. It is assumed that r_0 and K were constant over the years and over each monthly period. This is not true for K, but we may presume that any estimates obtained are some kind of average. We can assume that r_0 is really constant, however, and best estimated when N is far below K. Therefore, the value $r_0 = 1.6$ obtained in early spring is probably best and $K\dagger$ is the estimate of K when this value of r_0 is forced on the regression.

TABLE 2. *Thrips* population parameters

Time period	r_0	N_{max}*	N_{min}*	Correlation	K_0†
June–July	.015	26.9	0.8	−.45	7.0
July–August	.33	9.1	0.8	−.58	3.4
October–November	1.59	574	1.6	−.43	1,700
November–December	.85	574	3.6	−.80	177.3
January–February	−.58	37.8	2.9	−.74	9.14
February–March	.75	9.8	2.8	−.36	4.75

From Davidson and Andrewartha (1947b).
* Average number of *Thrips* per rose.
† With $r_0 = 1.6$.

As can be seen, except for the period from October to November, N varies about K, so the population during most of the year appears largely limited by K. In the early Australian spring, however, October K becomes much higher than N (1,700 versus an N maximum of 574) and it appears, as originally proposed by Davidson and Andrewartha, that spring *Thrips* populations are proximately not very limited by density-dependent processes in this season.

Aside

In the 1950s an earnest discussion ensued about the role of density-dependent versus density-independent factors in the regulation of populations. Andrewartha and Birch studied primarily insects and emphasized the importance of density-independent factors. David Lack, studying primarily birds, emphasized the necessity of density-dependent factors (see Orians, 1962). Many of the differences between their points of view seem related to the taxonomic group studied, but it was never very clear to me why insects should be regulated by one process while birds were regulated by another. The present theory has provided some insight into this matter which I shall now present.

It seems that most species in a seasonal environment are limited primarily by the winter season. That is, density-dependent factors operate most importantly when K is lowest. Rising populations are evidently so far from K that variations in $K - N$ have relatively little effect. This means that the population minimum (N_s) is regulated by density-dependent processes (usually in operation before N reaches N_s) while the population *maximum* (N_w) is regulated by density-independent processes. Now, insects are chiefly noticed at the peaks of their cycles (N_w), when they are the greatest nuisance. But birds, which are appreciated chiefly for esthetic reasons, are most interesting in spring, when their populations are lowest (N_s). Thus, bird students are apt to be concerned about N_s, a density-regulated figure, while entomologists are apt to be concerned about N_w, which is not so much density-regulated. Presumably, if Lack had studied fall populations of birds while Andrewartha and Birch studied prebreeding populations of insects, they might have taken each other's point of view. Lack might have discovered that the population growth of fledged nestlings depends primarily on density independent factors (spring and winter temperatures?). Andrewartha and Birch might have claimed that competition and K were all one needed to understand the number of *Thrips* in July, before the spring rains came.

The above analysis is crude, so the following refinement is suggested. This refinement depends on an analysis of original and very good data. No example is given here.

We have again the differential equation:

$$\frac{dN}{dt} = N' = r_0\left(N - \frac{N^2}{K}\right), \tag{2}$$

and we wish to estimate r_0 and K. We shall assume that r_0 is constant, but we shall let K be a function of time. Given a set of data, we can fit a polynomial of N versus time, for example,

$$N = B_0 + B_1 t + B_2 t^2 + B_3 t^3 + B_4 t^4 \tag{3}$$

with first derivative

$$N' = B_1 + 2B_2 t + 3B_3 t^2 + 4B_4 t^3. \tag{4}$$

At a given instant of time, t_1, we know N from (3), and from (4), N'. These known values can be substituted into (2) to give us an equation in r_0, the intrinsic rate of natural increase, and $K(t)$, the value of K at time t.

One equation in r_0 and K is of value since if one of these two should be known, the other can be computed. Also, if we could find another equation in these unknowns, independent of the first, we could solve for both unknowns. In order to pursue the latter possibility, let us consider N'' and N'''. In calculating N'' and N''', we obtain from (2)

$$N'' = r_0 \left(N' - \frac{2NN'}{K} + \frac{N^2K'}{K^2} \right) \tag{5}$$

and

$$N''' = r_0 \left[N'' - \frac{2N'^2}{K} - \frac{2NN''}{K} + \frac{4NN'K'}{K^2} + \frac{N^2K''}{K^2} - \frac{2N^2K'^2}{K^3} \right]. \tag{6}$$

Also, from (4)

$$N'' = 2B_2 + 6B_3t + 12B_4t^2 \tag{7}$$

and

$$N''' = 6B_3 + 24B_4t. \tag{8}$$

Remember that K is a function of time. Let us assume that it is a linear function. This assumption will be approximately met over any short time interval (region). Then K' will be constant and $K'' = 0$. If $K'' = 0$, (6) can be written:

$$N''' = r_0 \left[N'' - \frac{2N'^2}{K} - \frac{2NN''}{K} + \frac{4NN'K'}{K^2} - \frac{2N^2K'^2}{K^3} \right]. \tag{6'}$$

Since N'' and N''' can be estimated from (7) and (8), which come from data equations, (2), (5), and (6′) are three equations with only three unknowns: r_0, K, and K'. They can therefore be solved for these unknowns. In fact, they yield

$$K^2[A] + K[B] + C = 0 \tag{9}$$

where

$$A = \frac{2[N''N - N'^2]^2(N')}{(N'N)^2},$$

$$B = \frac{2[N''N - N'^2][N''N^2 + 2NN'^2]}{N'N^2} - N'N''$$

$$- NN''' - \frac{4N'[N''N - N'^2]}{N},$$

$$C = \frac{2N''N^2 + 2NN'^2}{N'N^2} + 2N'^3 + 2NN'N''$$

$$- N'''N^2 - \frac{4[N''N^2 + 2NN'^2]N'}{N}.$$

Equation (9) can easily be solved for K, but yields two values. These values can be used to calculate K' and r_0. From (5) and (2),

$$K' = \left(\frac{K}{N}\right)^2 \left[\frac{N''(NK - N^2)}{N'K} - N'\left(1 - \frac{2N}{K}\right)\right]$$

$$r_0 = \frac{N'K}{NK - N^2}.$$

This analysis can be conducted for any time point t, so K' and K can be estimated at every time. There are some restrictions, however. The weakest assumption in the analysis is that $K'' = 0$. This restricts the *entire* analysis (including estimation of the polynomial relating N to time) to regions where changes in K are nearly linear. There is no difficulty in this, with modern computers, if the data are good enough to permit accurate estimation of the higher order coefficients in such a region. Also, the region cannot be so short that growth effects due to age-distribution changes affect the estimation of the polynomial. The analysis thus should be conducted over a region longer than the generation time of the organism. There must, of course, be many data points in such a region.

In actually applying this analysis, the data would be handled as follows. One can divide the population growth cycles into regions where it is expected that K is changing nearly linearly. Then one can fit a fourth order polynomial in time to the population values in just that period. The N, N', N'', and N''' can then be estimated at a series of points throughout the period, and two values of K, K', and r_0 estimated at each one of these points. The values should fall into two groups, each internally consistent. One of these groups should be somehow unrealistic (e.g., r_0 extremely high and variable) and can be discarded.

A check on the assumptions is always possible in that K' and K are both estimated. If $K'' = 0$, then estimates of K at a pair of points in time should differ by $K_1/\Delta t$, where K_1 is the estimate of K' for the earlier of the pair of points and Δt is the difference in time between the estimates. One should also plot the polynomial with the actual data points so that regions of poor fit to the data can be avoided in the analysis.

The Side-Blotched Lizard (*Uta stansburiana*)

Tinkle's (1967) recent study of the side-blotched lizard seems to be the most complete available on a member of the herpetofauna, and we can easily apply the present analysis to his data. This is accomplished in Figure 27, from which it is apparent that the (average) populations of this lizard are limited by breeding resources, the resources (or predators) limiting the number of young Utas that hatch from eggs. It is not clear what this factor is. It is possible that competition for food among the adult females while they are forming eggs limits the production of eggs. It is also possible that clutch mortality is density-dependent.

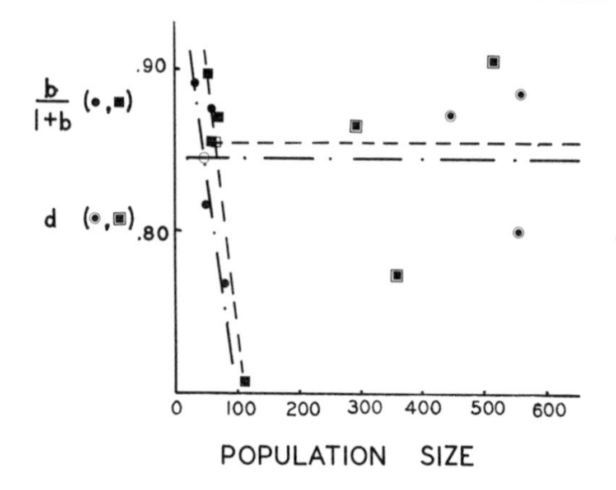

FIGURE 27. Population regulation in *Uta*. The broadly dashed lines represent one population (A), the finely dashed lines another (B). The equilibria are given as open points, the actual data as labeled. There appears to be no density dependence in d, but a great deal in $b/(1+b)$. The slopes for $b/(1+b)$ re N are, A, $g' = -.0034$ and B, $g' = -.0024$. To see whether these slopes are sufficiently steep that the winter population cannot be limited by winter resources, these slopes are to be compared with $1/(N(1+b)_{eq}$ (see Appendix to Chapter 1). The values are, for A, .0027 (less than $-g_A'$) and for B, .0020 (less than $-g_B'$), so both the A and B transformations are folded back.

Other Populations: Some Qualitative Remarks

There is a great deal of population literature to which the present theory might apply quantitatively. In these cases, as the above examples show, this can be rewarding. In many other cases, however, we have to be satisfied with only a qualitative application. The populations of aphids studied by Hughes (1963) may be taken as an example. Although the population involved is clearly under seasonal stress, it defies analysis by the simple ideas given above. Some of these population studies can be reviewed profitably in the present context.

Some of Nicholson's (1957) classic blowfly experiments contained one feature highly relevant to the present assumptions. In these experiments, he varied in a cyclical way the resources limiting the blowfly population. This, in effect, created a seasonal cycle in *K*. The effect of this cycling was obscured by internal competition cycles in the population, and evidently by evolution as well, so it is difficult to evaluate the results with regard to seasonal population problems in general. But the technique seems promising.

The Life Systems of Carpocapsa pomonella, the Codling Moth (after Clark et al., 1967)

Codling moths are an apple orchard pest. Adults lay eggs near or on maturing apples. The eggs hatch and the young larvae enter into the fruit and develop there. Nearing maturity the larvae leave the fruit and seek a place to pupate or diapause, depending on the season.

Thus there are several stages to the life cycle of this insect: adult, egg, larva entering apple, larva leaving apple and finding a place to pupate and pupa in the hibernaculum. There are several ways, therefore, that this species' population can be regulated. In fact, populations in different geographic regions appear to be regulated differently. Density dependence in fitness appears largely at two points in the life cycle: when larvae are seeking out apples and when larvae are seeking out places to pupate. The latter time is evidently consistently limiting, so the population in spring may be largely determined by the numbers of good hibernacula. During the growing season this spring population grows, and in areas with a long growing season it often becomes large enough to be limited by the availability of apples. In Nova Scotia, with a short growing season, the population never gets this high and so might be completely regulated by winter hiding places.

This is a rather remarkable possibility. Certainly the time spent as a pupa or in diapause seems ecologically uninteresting, at least with regard to energy flow. Yet if codling moths are limited by winter hibernation, then competition for hibernating sites, both intraspecific and interspecific, would be of prime importance in understanding the species' role in the community.

Of course, the codling moth is not by itself so important. But there are an enormous number of other animals, including many vertebrates (especially salamanders and other amphibians), which also hibernate over the winter months whose populations, by analogy, could also be regulated by competition for hibernating sites. This would, I think, make a considerable difference in the way we now view certain ecosystems.

We are interested in general patterns of population regulation. Specifically when there are several seasons or life stages, is only one limiting? So far, our examples have suggested this, but recently, Kiritani et al. (1970) and Kuno et al. (1970) have found that in leafhoppers, several of the life stages may be density-dependent, and that environmental variation does, in fact, move the population from one equilibrium to another, as we have predicted.

Plants

I have very little feeling for how useful the preceding discussion will be toward furthering understanding of plant populations. The present theories have all been developed around my own experience (actual, and in reading), which has not included a great deal of plant population ecology. I would not, however, exclude plant populations from possible consideration. Plant demography does not appear to have received the attention it deserves. Perhaps an attempt to fit the present theories to certain plant populations would be a worthwhile exercise.

SUMMARY

The direct application of the models is to analysis of population size. However, some of the conclusions from the analysis of the models have indirect importance elsewhere. Competition studies can be conducted only during those seasons when birth and death are density-dependent. Also the way in which a species is adapted may depend on which season limits its population.

Direct application of the long-generation models to the great tit in Marley Wood suggests that this species is limited by winter food. A similar analysis of Errington's bob-white quail data is more ambiguous, either because of curtailment of reproduction in anticipation of a bad winter or because of an unfortunate choice of study dates. Errington's muskrat data also suggest that the way in which Errington chose to take his data precludes a separation of seasonal effects, generally suggesting that prewinter survival is the regulatory phenomenon. The muskrat data also indicate habitat differences in population-regulating mechanisms.

A reanalysis of Davidson and Andrewartha's *Thrips* data using the short-generation models tends to support their original view of how that population is regulated. The spring increase in K seems to exceed by far the population's capacity to grow, with the result that usually the population simply increases as fast as it can as long as spring lasts. In the analysis of short-generation species data, good estimates of r_0, K, and K' (K' is dK/dt, the rate of change in K with time) at a given point in time can be estimated using good data fitted to a high degree polynomial.

Tinkle's (1967) data on *Uta stansburiana,* a lizard, when analyzed with long-generation time models, indicate that the population is regulated by density-dependent factors operating either on the laying female or the unhatched eggs.

There is a need for laboratory population studies where K is cyclically varied through time and for field population studies of the density-dependent survival of diapausing or resistant eggs or hibernating life stages.

CHAPTER FOUR

Some Analytical Models of Long-Generation Species

In the preceding chapter a number of data analyses were given that resulted in some hypothetical curves of population birth and death rates as these varied with population density. These curves were usually drawn by eye, which is never a very satisfying activity. It requires, however, a great deal of understanding of a particular system before models appropriate to this system can be proposed and fitted to the data. In this chapter, I shall offer two examples of such models for the great tit data.

A MODEL FOR WINTER MORTALITY

Observations

1. One of the more remarkable aspects of the great tit data is that the density-dependent mortality all seems to occur early in the winter, even before the beechmast falls, which apparently affects the d-$N(1 + b)$ curve.
2. The winter mortality after December seems to show no density-dependence whatsoever (Figure 19).
3. The apparent importance of the beechmast implies that winter food is limited to the great tit.
4. The hypothetical curves in Figure 18 suggest that there exists a minimal density-independent death rate, which determines mortality below some particular density.

Putting these suggestions into formal assumptions yields the following model:

Let the available food be N bird-winters, expressed in

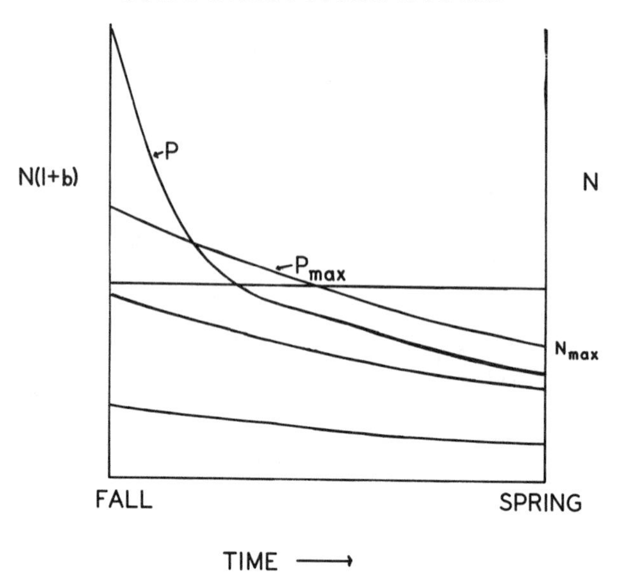

FIGURE 28. Model of nonbreeding survival when food is limited (see text).

Figure 28 as the rectangle. (One bird-winter is defined as enough food for one bird for one winter.) Assume this food to be fixed in abundance. Let the inevitable winter mortality be d_0, and assume that this mortality occurs uniformly over the whole winter (i.e., the instantaneous rate is constant) when the density falls below some level.

The average in Figure 19 may be regarded as an estimate of d_0. Then some possible curves showing the overwinter change in populations suffering only inevitable mortality are given in Figure 28 (P_{max} and unlabeled curves). The area under these curves is the food consumed by the populations, assuming (for simplicity) that there is no appreciable or systematic seasonal change in food requirements per bird over the winter. The upper curve denoted P_{max} inscribes an area equal to that in the rectangle, so this population eats all the available food. If the initial size of a population, P, is *higher* than the initial size of P_{max}, then the final size of P will be lower than or equal to the final size of P_{max}. This is true because, food being limited, the area under the P curve can be no greater than the area under P_{max}, and because P cannot have a lower mortality rate than the inevitable value. The latter restriction implies that P curves can only have a slope lower (more negative) than P_{max}, or for that matter than any of the inevitable mortality population curves in Figure 28. Then P curves cannot cross P_{max} or any of the unlabeled curves from below. So, if a P curve starts out higher than P_{max} but contains no more area, it must somewhere fall below P_{max} and must therefore cross it. Since it cannot cross back from below, it must stay below P_{max} and must terminate at a lower value. Therefore, P_{max} represents the maximum surviving population.

If a population does start out higher than P_{max}, it will most closely approximate the maximum surviving population if all the excess individuals die off at once. This seems to be what occurs in the great tit. In fact, the mortality of the excess in the great tit seems to occur *before* winter even begins (i.e., before the beechmast falls). Wynne-Edwards (1962) has developed the hypothesis that dominance hierarchies and territoriality are the mechanisms affecting this sort of mortality. I have elsewhere (Fretwell, 1969a) presented evidence supporting his views. Although the great tit has not been studied relative to this hypothesis, there are several reasons to believe that it might apply. On the negative side, the correlation of this mortality with the beechmast, even though the mortality occurs before the beechmast falls, is difficult to account for otherwise (Lack, 1966). On the positive side, late summer and autumnal expressions of territorial behavior are common and otherwise unexplained in this species (Hinde, 1952). Winter fighting for food is equally common (Lack, 1954).

We can formally model the above hypothesis as follows: Denote the prewinter population for P_{max} as N_w'. Assume that if the prewinter population $N_w = N(1 + b)$ is less than N_w', the mortality is d_0. If N is greater than N_w', then dominance (or something else) causes immediate mortality of the excess. Suppose in this case the initial value is denoted $N(1 + b)$. The final value for P_{max} can be denoted N_{max}, so the survival rate, S, is $N_{max}/N(1 + b)$. However, $S = (1 - d)$, so $d = 1 - N_{max}/N(1 + b)$.

N_{max} is the equilibrium breeding population while $N(1 + b)$ is a variable prewintering population. It is not surprising that the derivative of d in this model with respect to $N(1 + b)$ is $N_{max}/(N^2(1 + b)^2)$. At $N = N_{max}$ or equilibrium, this is $1/(N_{max}(1 + b)^2)$, which is exactly the value at which the breeding population ceases to be limited by the breeding resources (see Chapter 1). In the data for the great tit, N_{max} is seen to be a function of the beechnut crop and the year separation. Accepting a linear model of the dependence of N_{max} on the beechmast, the data in Table 1 lead to the following model for the great tit:

$$d = 1 - \frac{45.5 + 1.98 \text{ BM}}{N(1 + b)}$$

where BM is the beechmast rank as used in Lack (1966). This model applies only to pre-1956 data, for $N(1 + b) > N_{max}/d_0$.

Without dominance or some other controls, the above derivative would be even higher as long as the population is limited by winter resources. The excess prewinter population ($P - P_{max}$ at the beginning of winter) consumes food in proportion to its size and its period of survival. Then the food consumed by this excess implies a reduced surviving population. In Figure 28, the area beneath the P curve and the P_{max} curve before they cross represents this waste consumption which must be balanced by an equal area between the P_{max} curve and the P curve after they cross, if all the food is consumed. Increasing the latter area will in any simple example reduce the surviving population. Thus, lacking immediate mortality of the excess, an increase in $N(1 + b)$ will increase the area between P and P_{max} before they cross. In turn, it will increase the area between P and P_{max} after they cross, and will therefore result in a lower N. The derivative of d *re* $N(1 + b)$ is increased by the amount $(dN/dN(1 + b))(1/N(1 + b))$, and the d curve transformation folds back.

A MODEL FOR REPRODUCTIVE SUCCESS

We shall now develop a model for breeding success of the great tit. We can suppose that there is again a fixed amount of food but that one pair while feeding young can find only so much, even when there is no competition. The food that is found is largely fed to young in a nest and has to be carried to a central spot. The area around the nest which is searched will be determined by a balance between the depletion of insects about the nest by the feeding adults and the cost of flying a greater distance to undepleted feeding areas. Assuming a uniform distribution of food, the tit (without competition) wishing to maximize its catch will feed most intensely about its nest and less intensely elsewhere. At some distance from the nest (D) the cost in time as well as energy of flying to an undepleted searching area, when added to the actual cost of the search, will be greater than the cost of flying to and searching in a closer but somewhat depleted source. When the density of birds is low enough, the distance between nests is, on the average, approximately equal to $2D$. In this case the birds will never search the same areas, and therefore, cannot compete for food. We denote this density as $1/a_0$, where a_0 is the area available to each pair. For pair densities up to $1/a_0$, there is no competition and reproduction is maximal. Above $1/a_0$, competition occurs, but there is also a greater number of birds searching so more insects are consumed by the population. Let the total density of available insects (expressed in potential independent offspring per unit area) be n. Let the probability of any insect's capture at bird density $1/a_0$ be P. Then its chance of survival is $1 - P$. Assume that doubling the bird density simply doubles the search effort per unit area so the insect's chance of survival becomes $(1 - P)^2$. Actually, doubling the number of searchers may lead to competition, with a lower rate of insect harvesting per bird. If the birds were not already searching as hard as possible, they might respond to this reduction in yield with more searching, so an insect's chance of survival would be reduced even more. In general, then, raising the density to $1/a$ ($a < a_0$) lowers the probability of an insect's survival to $(1 - P)^{a_0/a}$. Thus, *if all the insects are identical,* $n(1 - P)^{a_0/a}$ will survive at a density of $1/a$ birds per unit area and $n(1 - (1 - P)^{a_0/a})$ will be caught. Each pair of birds has "a" units of land and thus receives $an(1 - (1 - P)^{a_0/a})$ fledged young units of insects. Assuming that maintenance of an adult is equivalent to producing one fledged young (for simplicity), this yields

$$2(1 + b) = an(1 - (1 - P)^{a_0/a})$$

$$b = \tfrac{1}{2}an(1 - (1 - P)^{a_0/a}) - 1.$$

If $A =$ the area of the habitat and $N =$ the number of birds, then $2/a = N/A$ and we have

$$\frac{a_0}{a} = \frac{a_0 N}{2A}.$$

Then,

$$\frac{b}{1 + b} = g = \frac{(An/N)[1 - (1 - P)^{a_0 N/2A}] - 1}{(An/N)[1 - (1 - P)^{a_0 N/2A}]}.$$

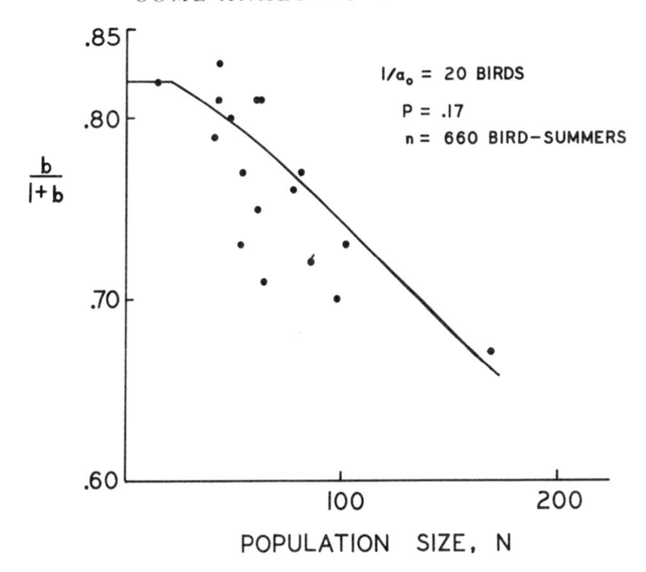

FIGURE 29. Application of model of reproductive success to great tit data (see text).

When this model is applied to the data, a fair fit can be obtained (Figure 29). There is some difficulty, however. It appears that at the higher bird population densities most of the available insects in the wood are caught and eaten (537 bird-summers out of 660). Also it appears that competition begins to occur at the very low density of about .5 pair/hectare (over 2 hectares, or about 5 acres, per pair). Finally, it appears that the data are generated by some process that at intermediate densities is more density-dependent than the theoretical curves of the model. These difficulties may be due in part to several factors. These are (1) the great and blue tit (*Parus caerulus*) fluctuate in abundance together and feed on similar foods in the breeding season. Perhaps they ought to be grouped together in the determination of population size. (2) There are other density-dependent factors affecting reproduction. For example, renesting fluctuates with density (Kluyver, 1951). (3)

Food supply probably varies from year to year in some regular way, and the trends in the change in food supply may accidently be correlated with the trends in change in population size. (4) The availability of food may really be correlated with the size of the population, either through direct effects of more intensive winter or early spring feeding on later population growth in the insects or through indirect effects, the insects being more or less abundant after a mast year. (5) All insect resources are not identical; some may be more easily caught than others. These objections are instructive, indicating questions that need to be answered before a more accurate model of reproduction in the great tit can be constructed.

Before concluding this section, a word of caution with regard to the particular models developed here is advisable. The data, despite their abundance, are insufficient to permit very confident acceptance or rejection of the developed models. I suspect that the model of winter mortality will prove much too simple when enough data are available to sort out the effects of temperature, predators, and catastrophes. The "inevitable" mortality rate probably varies slightly with density. Objections to the model of breeding production have been stated above. However, the models, while decidedly preliminary, are useful in that they suggest currently unobserved relationships. For example, the preciseness of control suggested by the winter food can be tested by experimentally providing winter food supplies to observe the effect on the population. The theory also predicts that dominance activity in the late summer should be correlated with the developing beechmast.

RESOURCE QUALITY AND UTILIZATION

In the theory of long-generation species in Chapter 1, several kinds of curves appeared. There were flat curves, sigmoid curves, curves with nonfolded transformations, and curves with folded transformations. Some general but brief discussion of what kinds of phenomena generate these different curves is given in this section. It may be possible to make *a priori* observations about population regulation from such considerations.

Flat Curves

Flat curves indicate an absence of density-dependence. This has two implications. First, it means that the members of the population are not influencing each others' fitness. As in the great tit models above, there can be some density below which survival or reproduction rates are simply determined by the organism against the environment. If the population never reaches this density, a flat curve will result.

The second implication is that the population is not influencing the available supply of its limiting resources, *relative to the utilization capabilities* of the individual organisms. If standing crops are reduced by such a population, it makes no difference to the organisms (see Rosenzweig and MacArthur, 1963, for another view of this sort of system).

Sigmoid Curves

Sigmoid curves can be thought of generally as approximations of a combination of two other curves, one flat and one density-dependent. It seems likely that when any species is rare enough, it ceases to make much of an impact on either itself or its environment. Thus the relationship between birth and death rates should be flat at low population sizes. A sigmoid curve, however, ignores Allee-type effects where g increases or d decreases with respect to N at low densities. Also ignored are the problems of extreme rarity. We assume that these problems are confined to an uninteresting part of the curve. There is a theoretical reason for supposing this; there can be no stable equilibrium in the "too rare" region of such curves. However, these curves are interesting on other accounts, some of which will be discussed later.

The curved part of a sigmoid-type curve can be generated by a variety of models which reflect either interference of organisms with each other or depletion of environmental resources or stimulation of predators.

Transformations

The transformation curves describe the population at the end of a season as a function of the birth or death rates at the beginning of that season. There are two kinds of transformations, folded and nonfolded. If a curve has a nonfolded transformation, it implies that resources cannot be overexploited. A folded transformation implies that overexploitation can result at certain population sizes.

MacArthur (pers. comm.) has recommended as an example of the first kind of resources the case of a waterthrush feeding on a tidal mangrove. Orians and Horn (1969) have provided another example, blackbirds feeding on emerging odonata. The food for such a bird might be replaced once or twice daily, so resource density on the average is little influenced by bird-feeding activity, except for very short periods. Thus, if there was density-dependence in such a system, it might arise from different individuals trying to live in an area too small to support them all. Some inefficient individuals might persist over a long period before finally dying in such a system, but the food supply, after they have died, will not be affected by their persistence.

Orians (1969) points out that blackbirds feeding on emerging odonata are similarly disposed, the birds' utilization having little short-term effect on the population of resources. This example is contrasted with the organism that depends on a "storehouse" full of resources that can easily be depleted (like the beechmast of the winter great tit model above). Unless it has very effective dominance regulation, the population suffers permanently from free competition. The longer poor competitors are allowed to persist, the more food these doomed individuals consume and the lower the ultimate surviving population. These "storehouse" populations (which, of course, are a theoretical idealization) have the additional problem that efficiency cannot alone discriminate effectively between competitors; when the storehouse is first filled, there is too much resource for efficiency to figure significantly. However, if the storehouse has almost enough resources for the whole population, more efficient consumers will have an advantage. Towards the end of winter, as resources become low, the more efficient individuals can develop within themselves enough reserves to survive the rest of the winter.

Wynne-Edwards' (1962) hypothesis, in the present theoretical frame, probably concerns folded transformations. One might say that Wynne-Edwards' theory asserts that organisms generally have folded transformations, but use social behavior to straighten up their transformations. That is, from Wynne-Edwards' ideas, we might predict that most species will actually have vertical transformations in the limiting season(s), and that the transformations would be folded but for the existence of certain social mechanisms. Mechanisms for the evolution of such social control may well involve, as originally asserted by Wynne-Edwards, selection at the level of self-replicating groups, as long as these groups are sufficiently inbred and consist largely of kin organisms (Smith, 1964).

SUMMARY

The graphical curves in the long-generation models of the preceding chapters were given only the crudest logical justification. More precisely developed models are necessary for a real understanding of population regulation, and examples of such models for the great tit in Marley Wood can be developed.

By supposing that wintering great tits are limited by food, a theoretical maximum surviving population can be defined, given some density-independent maximum survival rate $(1 - d_0)$. Populations which begin the winter at a larger size than the theoretical maximum must overeat their food resources, and the surviving population will be lower. Perhaps dominance or social behavior regulates the initial population through epideictic displays so that the surviving population is maximized. Then a nonbreeding season death rate fall population size model is

$$d = d_0 \qquad \text{for } N(1 + b) < N(1 + b)_0$$

$$d = 1 - (1 - d_0)\frac{[N(1 + b)]_0}{N(1 + b)} \qquad \text{for } N(1 + b) \geqslant [N(1 + b)]_0$$

where d is the death rate over the entire winter, d_0 is the minimum death rate, $(1 - d_0)[N(1 + b)]_0$ is the maximum surviving population, and $N(1 + b)$ is the fall population.

For the great tit in Marley Wood, $(1 - d_0)[N(1 + b)]_0$ is a function of the beechmast, estimated as 45.5 + 1.98 BM (BM is a beechmast rank).

A bird gathering food for its young in the breeding season is confronted with a large supply of food that is hard to find. Thus, a higher density of birds represents more searchers, so more of the available insects will be found. There may be some overlap of search, however, so the per pair catch may be diminished at higher densities of searchers. Then the per pair production of young will be density-dependent. A model constructed from these ideas can be applied to the reproduction data of the great tit. The fit is not as good as might be expected; the analysis suggests that almost all the available insects are caught at the higher tit densities. Also, there seems to be more density-dependence at intermediate densities than this model allows.

These models are specific. However, it is possible to make some more general remarks about the nature of the differences between curves. Flat curves indicate that individuals of a population affect neither each other nor their environment in any meaningful way. Sigmoid curves which are initially flat but show density dependence at higher population sizes are combinations of two effects. A population need not affect itself as much when rare as when it is more abundant. Curves with nonfolded transformations arise from populations which feed on resources that cannot be overexploited. A folded transformation indicates the possibility of overexploitation.

SECTION II

HABITAT DENSITIES

CHAPTER FIVE

Theory of Habitat Distribution

Organisms living in a seasonal environment frequently disperse or migrate in order to avoid or better endure periodic hardships. If a population lacks such movements, its distribution is a simple function of its abundance (Andrewartha and Birch, 1954) and is an interesting problem only in regard to abundance itself. In effect, in sedentary populations, the theory of habitat distribution is simply a list of theories of abundance in all available habitats.

For dispersive populations, however, it may not be possible to develop theories of abundance for separate habitats. The individuals in a habitat may be numerically affected by events in other habitats and by events prior to dispersal to the habitat. The analysis of distribution in such cases is not a simple problem. It demands some kind of solution as "dispersive organisms" includes almost all animals as well as many plants.

In this section, Chapter 5 will develop some simple ideas about the habitat distribution of dispersive species, and will present some brief examples applying these ideas. I have made a more thorough effort to apply the basic theoretical ideas to breeding passerine bird populations; the results of these studies will be presented as Chapter 6. The winter distribution and abundance of some of these passerines will be considered at greater length in Chapter 6. The orientation of these chapters is mostly toward avian populations and communities. This represents, I think, the decline of generality in the face of increased realism

(Levins, 1966). It also reflects the limited experience of the author. In any case, it may suggest applications in other taxonomic groups, and I look forward to a more general development.

HABITAT DISTRIBUTION

The following theory of habitat distribution will be developed about the idea of unspecified curves of density-dependent fitness. We first define what is implied by the terms "habitat" and "habitat distribution," following the arguments of Fretwell and Lucas (1970).

Definitions

Habitat. A habitat of a species is any portion of the surface of the earth [1] where the species is able to colonize and live (temporarily *or* permanently) at some density greater than zero. The total area available to a species can be divided into different habitats. The area of any one habitat can be large or small, and different habitats of the same species may be of different sizes. A given habitat can consist of several subdivisions which are not contiguous. We shall define habitats so that all of the area within each habitat is, at zero density of the species, essentially homogeneous with respect to the physical and biological features which we believe to be more relevant to the behavior and survival of the species. We shall also frame our definitions so that different habitats are not identical with respect to those same physical and biological features.

This definition does not imply that all measurable variables within a single habitat must take constant values over all of that habitat. Some variables may be irrelevant. Others, such as temperature and humidity, may compensate for

[1] In a number of cases, a habitat may be restricted to a layer parallel to the surface of the earth.

one another. In the latter case, the "relevant feature" which is "homogeneous" is some function of the compensating variables. Thus, evaporative heat-loss rate, which depends on temperature and humidity, may be the relevant feature that is homogeneous, in which case temperature and humidity may still vary over a single habitat. Because of compensating variables, all of the area within a habitat does not need to *appear* homogeneous to our measuring devices. However, if an area is uniform in all measurable variables, it is in a single habitat. Thus, all space within a "homogeneous" habitat has one property and only one in common: The fitness of any random phenotype is identical at all points within it.

Habitat distribution of a species. Suppose the total area available to a species is divided into different habitats and that the area of each habitat is known. The habitat distribution of the species is the set of numbers which state the number of individuals resident in each of the habitats. It can also be expressed as the proportions of the total population resident in the different habitats or as the density in each habitat.

Factors Affecting Habitat Distribution

Habitat selection. We can now consider how the habitat distribution is achieved. Habitat distribution in dispersive organisms may, in part, be based on habitat selection, at least some individuals being exposed to a variety of habitats of which just one is chosen for residence. Therefore, the distribution may be considered as a behavioral phenomenon involving stimuli and responses. This means that an understanding of the habitat selection responses in given environmental circumstances will lead to an understanding of the habitat distribution. In order to understand behavioral responses, we should consider first the environmental factors (excluding direct within-species individual interactions) which caused the natural selection leading to the evolution of the behavior.

Suitability. In the case of habitat choice, these factors include differences in goodness or suitability of habitat because individuals which choose relatively poor habitats are selected against. Although the stimuli directly influencing the choice of habitat may be no more than correlated with habitat goodness, it is the goodness itself which is a basic (or ultimate—see below) determinant of the behavior.

To summarize, the relative suitabilities of the different habitats give rise, through evolution, to habitat selection which in turn determines the habitat distribution. The habitat distribution then depends on the relative goodness of the habitats.

In developing a theory of habitat distribution, the next matter is to examine the relative suitability of the various habitats. Suppose the habitats are indexed i, $i = 1, 2, \ldots, I$, where I is the total number of habitats. The goodness of each occupied habitat is related to the average potential contribution from that habitat to the gene pool of succeeding generations of the species. We are interested in some measure of that goodness, which may be called the suitability and denoted for the ith habitat as S_i. The suitability of the habitat cannot here be defined precisely, but may be thought of as the average success rate in the context of evolution (and/or "adaptedness") of adults resident in the habitat. Stated formally, if S_{iq} is the expected success rate of the qth individual ($q = 1, 2, \ldots, n_i$ where n_i = number of birds resident in the ith habitat), then

$$S_i = \frac{1}{n_i} \sum_{q=1}^{n_i} S_{iq}. \tag{10}$$

The habitat suitability will be determined by several factors such as food supply and predators. The influence of some of these factors is density-dependent, so the suitability in a habitat is affected by the density of birds there. Let us assume for the moment that the effect of density is always a decrease in suitability with an increase in density. This assumption would imply that Allee's principle does not operate (Allee et al., 1949); thus the assumption may not be valid when densities are close to zero. Allee's principle states that per capita survival and reproductive rates increase with population size up to some maximum. Further increase in population size leads to a decrease in survival and reproduction, as assumed here. Allee's principle certainly holds at very low densities. A solitary male, for example, cannot have as high a reproductive rate as a male-female pair. At moderate densities, the assumption that suitability decreases with increased density is reasonable since predators may become more active at higher densities and competition for food more severe. We shall return to Allee's principle later.

We can now define a habitat distribution which will provide a reference for the study of dispersive populations. This is the ideal free distribution. It rests on assumptions about habitat suitability and the adaptive state of organisms.

Ideal Free Distribution

Assumptions about suitability. Ignoring Allee's principle for the moment, if we assume that suitability always decreases with density, then it would follow that the maximum suitability occurs when the density approaches zero. Let us call this maximum value the *basic suitability,* denoted for the ith habitat as B_i. The basic suitability of the ith habitat is affected by such factors as potential predators, food density, and cover.

These considerations lead to an equation expressing the suitability of the ith habitat as a function of the basic suitability there and the density (denoted d_i). We write

$$S_i = B_i - f_i(n_i),\ i = 1, 2, \ldots, I. \tag{11}$$

The term $f_i(n_i)$ expresses the lowering effect on suitability of an increase in population density in the ith habitat. Since $f_i(n_i)$ always increases with density, S_i always decreases. Equations (11) will here be assumed to be the same through time. A possible example of Equations (11) for some value of i is plotted in Figure 30.

Before going any further, let us order the habitats in terms of their basic suitabilities for some particular phenotype so that $B_1 > B_2 > \cdots > B_I$. By definition, no two habitats have equal basic suitabilities. This is consistent with our restriction on habitat definition (stated above) that no two habitats are identical with respect to relevant features.

Assumptions on organisms. A description of the suitabilities of the various habitats has been considered in order to understand the habitat selection behavior and the habitat distribution. In applying this description, we make two additional assumptions. These are (1) that all individuals settle in the habitat most suitable to them and (2) that all indi-

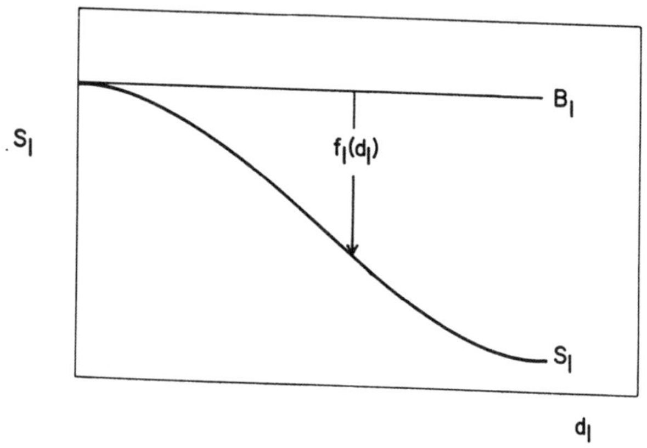

FIGURE 30. Suitability versus density: first habitat (see text). (From Fretwell and Lucas, 1970.)

viduals within a habitat have identical expected success rates.

The first assumption demands that the organisms have habitat selection behavior which is ideal in the sense that each selects the habitat best suited to its survival and reproduction. Such organisms will be referred to as *ideal* individuals. It is not an unreasonable assumption since individuals which are closest to being ideal will be selected for in the evolution of the species. Therefore, if the environment has sustained the same selective pressures for a large number of generations, the behavior of actual individuals should be approximately ideal.

The second assumption demands, first of all, that the individuals be *free* to enter any habitat on an equal basis with residents, socially or otherwise. For example, if a population in a habitat is limited by nest holes and if all these nest holes are occupied by residents which neither share nor are displaced, then a newly settling individual may expect to be totally unsuccessful, although the average of all residents is rather high. In this case, newly settling individuals are *not* free to enter the habitat on an equal basis with residents. If, when a new bird arrived, all of the occupants of the habitat came together to draw lots for the nesting holes and those losing remained in the habitat, then the individuals would be free. This, of course, is unrealistic. We make this sort of assumption to provide a basis from which reality may be more objectively considered. The assumption is certainly never realized but may be useful anyway. A more familiar example of this sort of analysis is the ideal gas law ($PV = nRT$), which is based in part on the assumption that gas molecules have neither mass nor volume. The invalidity of this assumption detracts very little from the usefulness of the result.

Later on, when we discuss territorial behavior, this assumption will be relaxed.

The second assumption demands also that individuals be alike, genetically and otherwise. This aspect of the assumption may restrict the application of this theory to local populations.

A particular difficulty which arises from this assumption concerns habitat accessibility. According to this assumption, if accessibility is relevant, every habitat must be equally accessible to all members of the species. This is absurd when we regard widely distributed species and only local habitats. However, if we restrict attention to habitats as widely distributed as the species or to local populations as narrowly distributed as the habitats, then this difficulty is bypassed.

Note that the second assumption and Equation (10) imply

$$S_i = s_i.$$

Each bird expects success equal to the habitat average.

The ideal free distribution. With these assumptions we now use the suitability Equations (11) to determine the habitat selection of the individuals in the population. The ideal assumption states that each individual will go where his chance of success is highest. The assumption of homogeneity states that each individual's chance of success is highest in the habitat of highest suitability. The two assumptions together then assert that each individual will go to the habitat of highest suitability. Thus, a description of the relative habitat suitabilities to some degree determines the choices of ideal free individuals. These choices in turn determine a distribution which may be called the *ideal free* distribution. This distribution is formally described below.

If all individuals choose the habitat of highest suitability, then from the point of view of unsettled individuals, the suitability in all occupied habitats must be approximately equal, and larger than or equal to the suitability in all unoccupied habitats. This is true because if some occupied habitats had a clearly lower suitability, then some of the birds in that habitat could improve their chance of success by moving to the habitats of higher suitability. If they did not make that move, they would not have ideally adapted habitat selection behavior, contradicting the ideality assumption. The distribution is stable only when suitabilities are equal in all habitats. With Equations (11), a fixed set of habitat areas, and given population size, the condition of equal suitabilities in all occupied habitats determines completely the ideal free distribution. To prove this, let a_i be the area of the ith habitat and M the population size. If exactly l habitats are occupied, then the equal suitabilities condition says that

$$S_1 = S_2 = \cdot \cdot \cdot = S_l. \tag{12}$$

Note that the first l habitats are the l occupied habitats. These are the l habitats with the highest basic suitabilities, B_i. Let us prove this. If the $(l + p)$th habitat were occupied, some $(l - q)$th habitat must be unoccupied (p,q positive integers such that $0 < p \leqslant I - l$ and $0 < q < l$), since only l habitats are occupied. But because n_{l-q} equals zero, the suitability in habitat $l - q$ is B_{l-q} so:

$$S_{l-q} = B_{l-q}.$$

Also, recall that $B_1 > B_2 > \cdot \cdot \cdot > B_l > B_l + 1 > \cdot \cdot \cdot > B_I$ by definition. Then

$$B_{l-q} > B_{l+p} \geqslant S_{l+p}$$

or

$$S_{l-q} > S_{l+p}. \tag{13}$$

Because, under our assumptions, individuals settle where the habitat suitability is highest, the birds in habitat $l + p$ would move to $l - q$ where by (13) the suitability is higher. Thus, if there are l occupied habitats, they are the first l.

It is also true that the total population size (denoted N) of a given species over all its occupied habitat is given by

$$N = a_1n_1 + a_2n_2 + \cdots + a_ln_l, \tag{14}$$

since a_in_i is the number of birds in the ith habitat and the total population is the sum of all the birds in all occupied habitats.

From Equations (11)

$$S_i = B_i - f_i(n_i),\ i = 1, \ldots, l,$$

so $S_i = S_{i+1}$ in (12) implies

$$B_i - f_i(n_i) = B_{i+1} - f_{i+1}(n_{i+1}),\ i = 1, \ldots, l-1. \tag{15}$$

There are $l-1$ of these equations in l unknowns; $n_1, n_2, \ldots, n_l$. These plus Equation (14) make l equations. These l equations can be solved uniquely[2] for the $n_i (i = 1, \ldots, l)$ in terms of the constants N and the a_i. The distribution can be expressed as the proportion of organisms in each habitat. Denote the proportion in the ith habitat as P_i. Then clearly $P_i = n_ia_i/N$, and the distribution is seen to be a function of the density in each habitat. Since the densities are determined by the condition of equal suitabilities, so is the distribution.

As example of the solution of the equations is given in Figure 31 for $l = 1, 2, 3$. The suitability curves are drawn for three habitats, 1, 2, and 3. When no animals are present, the suitability is highest in 1 and equals B_1. Therefore, if a small number now settle in the habitats, they will all go to 1 because they settle where the suitability is highest. Then the density in 1 will increase from 0 and the suitability will decrease, following the curve labeled S_1. As the population size increases, more and more individuals will settle in 1 until the density there is so high that the suita-

[2] Because the $f_i(n_i)$ are always increasing.

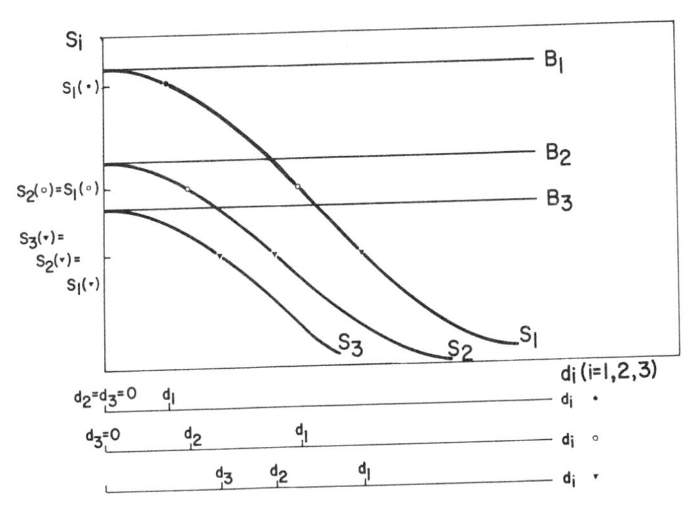

FIGURE 31. Suitability versus density: habitats 1, 2, and 3. The ideal free densities are shown on the extra density coordinates at three values for the total population size, $M' < M'' < M'''$. The situation at each population size is denoted by (●) for M', (○) for M'', (▼) for M'''. At M', the lowest population size, all the population is in 1; the densities in 2 and 3 are zero. At M''', the largest population size, all three habitats are occupied. (From Fretwell and Lucas, 1970.)

bility is equal to the basic suitability in 2 (B_2). Now, any additional organisms have a choice of habitats; 1 and 2 are equally suitable. However, these additional birds must increase the density in both habitats in such proportions that the suitabilities in both remain equal. Further increases in population size will raise the density in both habitats. If the population increases enough, the densities in 1 and 2 will be so high that the suitabilities in both habitats are reduced to B_3, the basic suitability in habitat 3. Any additional organisms must increase the density in all three habitats in such proportions that the suitabilities in all three remain equal.

Allee-type Ideal Free Distribution

Let us now briefly consider the effect on the distribution of Allee's principle, which we have heretofore assumed does not apply. In this case, the S_i curves first increase with density up to a maximum then decrease. These curves do not always have unique inverses; there are sometimes two densities corresponding to a single suitability. Therefore, Equations (12) and (14) do not necessarily have a unique solution. Consideration of Allee's principle suitability curves is best done graphically as in Figure 32. At low population sizes, the organisms will presumably go to habitat 1, and as the population increases, will enjoy an increasing suitability up to some maximum. Further increases in population size with all individuals settling in 1 will cause a decrease in suitability in habitat 1 until at some higher population size (A) the density (n_1) in 1 is such that the suitability there equals the suitability in 2 at density 0. Now a remarkable

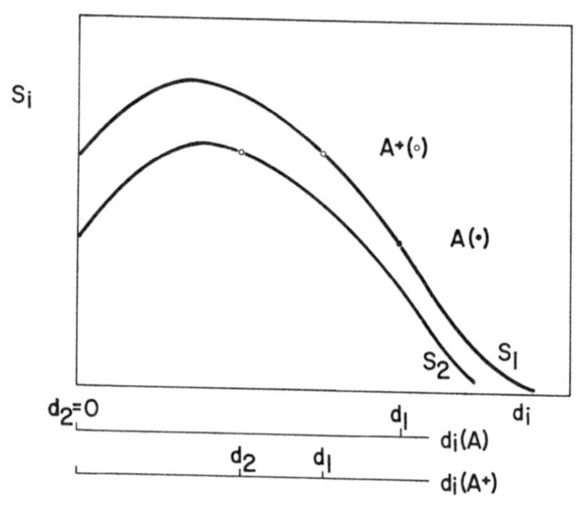

FIGURE 32. Suitability versus density under Allee's principle. At population size A the density in habitat 1 is $d_1(A)$. At population size $A+$ the density in 1 is $d_1(A+)$ while the density in 2 has increased to $d_2(A+)$. See text for explanation. (From Fretwell and Lucas, 1970.)

event may occur. With a further slight increase in population size ($A+$), some individuals will settle in 2 and perhaps some in 1. But the suitability in 2 *increases* with an increase in density while the suitability in 1 decreases; therefore, $S_1 < S_2$, and suddenly it becomes advantageous for individuals in 1 to go to 2. Being ideal, they will so move, and may continue to move until the suitability in 2 is maximal. They would then fill the two habitats in such a way that $S_1 = S_2$ and may well become common in both (open circles in Figure 32). Thus, a very small increase in population size may result in a very large change in the distribution. By manipulating the S_1 curves, many such changes can be produced. One can generally conclude that under the conditions of this theory, species following Allee's principle may demonstrate erratic changes in distribution with small changes in population. One can imagine curves which lead to complete shifts in population while other curves may lead to no erratic behavior at all.

Dominance and Territorial Behavior

We have so far specifically excluded dominance and territorial behavior in that interorganism effects on fitness were assumed not to exist. It is important, however, to consider the possible relationships such behavior might have on habitat distribution. Three hypotheses have been proposed about the role of territorial behavior in distribution; we consider all three.

We first make a general definition of territorial behavior. In a later section, the results of the territorial discussion will be related to dominance behavior.

Territorial Hypotheses and Effects on Distribution

Definition: Territorial Behavior

Territorial behavior has commonly been defined as site-dependent defensive behavior, the territory being "a defended area" (Noble, 1939). This definition is simple and attractive. However, I shall show later that it is really ambiguous and, therefore, altogether unsatisfactory. The historical problem of the function of territoriality may be traced back to just this sort of definition and the fact that it means different things to different people. In order to include most of the usual behavior normally called territoriality, I prefer to define the behavior as follows:

Territorial behavior is any site-dependent display behavior that results in *conspicuousness* and in *avoidance by other similarly behaving individuals.* Territorial behavior is specifically not restricted to defensive and/or aggressive behavior nor are they excluded.

The following hypotheses were in each case inspired by the authors cited in connection with them.

The Density Assessment Hypothesis

This first hypotheses ascribes a role to territoriality which permits achieving the ideal free distribution (non-Allee). It was first clearly described by Kluyver and Tinbergen (1953) and given fullest development by Wynne-Edwards (1962). Before discussing the hypothesis and its consequences in distribution, some preliminary discussion on the achieving of an ideal free distribution will be presented.

Given the total population size N, the habitat areas a_i, and the equations in S_i (11), the density in each habitat is determined. The density in turn determines the proportion of the population in each habitat or an ideal free distribution. The actual values will be expressed in terms of N and the a_i which for a given year are constants for the purpose of this discussion. However, N, and perhaps the a_i vary with time in a somewhat irregular fashion. Therefore, the proportion of the population in a given habitat may also vary between different years. This is demonstrated in Figure 31. The suitability curves for three habitats are drawn and the habitat densities at three population sizes shown. The relative densities in each of the three habitats are shown on the extra abscissas and are markedly different at the three population sizes.

If the total population size does not vary, then a species could consistently achieve an ideal free distribution by being composed of individuals of which a fixed proportion prefer each habitat at all times, or by being composed of individuals which prefer each habitat a fixed proportion of the time. However, if the variation in population size does occur and leads to considerable variation in the ideal free distribution, then the habitat selection of the individuals, if it is to be ideal, must adjust to the changing conditions.

In order for individuals to be able to modify their habitat selection in accordance with changes in the ideal free distribution, there must be some cue or cues which reflect these changes. The changes come about due to variation in population size and possibly habitat areas. These variables affect the densities in the various habitats and therefore the relative suitabilities. Any cue, such as footprint abundance, which would reflect habitat densities or any cues which would reflect population size and perhaps habitat areas could be used by ideal individuals in achieving the ideal free distribution. The individuals would have their preference for a given habitat depending on the state of the cue.

Kluyver and Tinbergen (1953) suggested that the territorial behavior of resident individuals is used as a density cue (or Wynne-Edwards, 1962; an "epideictic" display) by unsettled individuals so that they can avoid highly populated habitats where the chance of breeding success is presumably lower than elsewhere. These authors observed that the habitat distribution of some *Paridae* (tits) was dependent on population size. At low population levels, most of the individuals were found in a single habitat type, while at higher levels many individuals occupied another habitat type at a lower density (compare this with population sizes M' and M'' in Figure 31). They emphasized that breeding success (and therefore suitability) was not noticeably different in the two habitats. Thus, the distribution of the tits was apparently nearly an ideal free one despite changing population size. No appropriate cue for density other than territoriality was observed nor was there evidence for a cue for population size.

There is no evolutionary difficulty in supposing that territorial behavior serves as a density index. It is obviously to the residents' advantage to provide such a cue since they suffer if their habitat is crowded to the extent that their suitability is lower than in other habitats. It is also to the advantage of the settling individuals to respond to such a cue, for by so doing they avoid habitats where high density makes their chance of success lower than elsewhere. Since the population size of tits varies considerably, the development and use of some density cue, such as territorial behavior, might be expected.

Perhaps the most important aspect of the present model can now be realized. In the curves that are offered, there is clearly no sense in which a habitat can be "full." More birds may always settle in a given habitat. They may not do so because of reduced suitability, but the reduction need not be absolute to prevent a higher density. Organisms are never prevented from settling because habitats are "full." They may refuse to settle at one time because they intend to settle elsewhere later, either in the same season, or in long-lived organisms, in a later season (Rosenzweig, pers. comm.).

The Density-limiting Hypothesis

Introduction. The second territorial hypothesis is based on a model from Huxley (1934), who described a territory as a rubber disk. The disk can be compressed, but an increased amount of force is necessary as it gets smaller. This hypothesis is relevant to the free aspect of the homogeneity assumption in the ideal free model. That assumption states that any individual is free and may therefore enter any habitat on an equal basis with the individuals already resident there. With the rest of the homogeneity assumption, this means that the average success of the occupants of a habitat is also the suitability that an unsettled individual will have (on the average) on settling in that habitat. The free assumption, as already mentioned, fails if the species is limited by nest holes which, once occupied by a resident, are not relinquished or shared. The second territorial hypothesis describes another possible way in which this assumption might fail. Suppose the residents of the habitat, by their territorial behavior, make it dangerous for unsettled individuals to enter the habitat. Then the average success of newly settling individuals will be lower than the habitat average, and the assumption fails. If so, ideal individuals maximizing their own success will not necessarily settle where the habitat suitability is highest, and the habitat suitabilities no longer must be equal. Since Equation (15) no longer holds, the distribution is not determined as before. There exists a new, different distribution.

The supposition that the territorial behavior of residents restricts nonresidents from settling is reasonable. In evolution such behavior effectively prevents the density in the habitat from increasing, thus maintaining the suitability (see Figure 33). This would give the aggressive residents a selective advantage and the behavior, as suggested by Brown (1964), would spread throughout the population.

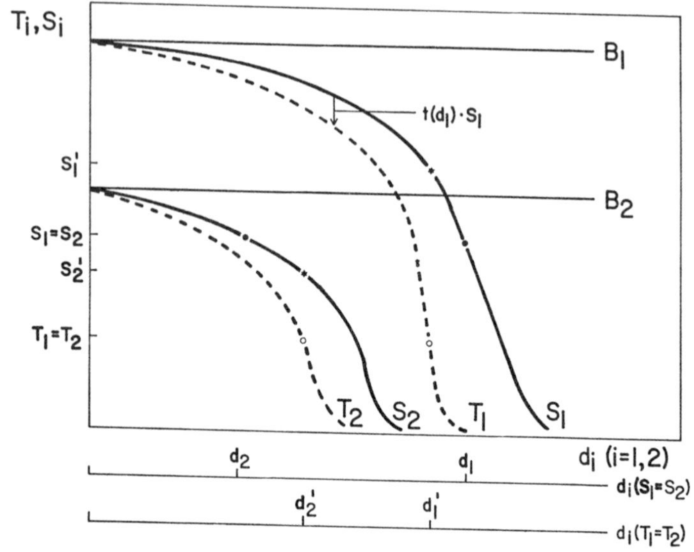

FIGURE 33. A comparison between ideal-free and ideal-despotic distributions: habitats 1 and 2. Population size and habitat areas are constant. S_i is the suitability for established individuals; T_i, the apparent suitability to unestablished individuals. The filled circles represent equilibrium conditions $S_1 = S_2$ for the ideal free distribution, and densities for that distribution are given below as (d_1, d_2) on the abscissa labeled $(S_1 = S_2)$. The open circles represent equilibrium conditions $T_1 = T_2$ for the despotic distribution, and densities for that distribution are given as (d_1', d_2') on the abscissa $(T_1 = T_2)$. At $T_1 = T_2$ the actual suitabilities are given as (S_1', S_2') marked on the S_i curves with a star. Note that in going from the ideal free to the ideal despotic distribution, at constant population size and habitat areas, the density in 1 decreases $(d_1' < d_1)$ while the density in 2 increases $(d_2' > d_2)$. Some individuals have in principle been forced from habitat 1 into habitat 2. See text for further discussion. (From Fretwell and Lucas, 1970.)

However, if all the individuals, settled and unsettled alike, became equally aggressive, it might not then be possible for the settled individuals to make a habitat less suitable to unestablished individuals. This is perhaps unlikely; for example, dominance seems to depend on experience (Nice, 1957; Sabine, 1955), which for a given area should vary considerably among organisms. In the following discussions we shall assume that all individuals are not equally aggressive so that social dominance hierarchies are established and the free assumption does not hold.

Consequence of hypothesis for the distribution. We consider in more detail the altered ideal distribution which would arise from this hypothesis. By definition the ideal bird always goes to the habitat where his potential success is highest. In the ideal free distribution, the potential success of a new individual settling in a given habitat is equal to the average of all individuals resident there, including the new one. This is the habitat suitability described by Equation (11). If the unsettled individuals are restricted by the territorial behavior of residents, then their potential success is less than the average of the habitat. This suggests that we define some quantity t, $0 \leq t \leq 1$, which can be subtracted from one and multiplied into the S_i to yield the apparent habitat suitability from the point of view of the unsettled organism. This apparent habitat suitability may be denoted by T_i, defined symbolically in Equation (16):

$$T_i = S_i(1 - t). \tag{16}$$

Let us now consider some of the properties of t. This quantity will depend on density and can be reasonably assumed always to increase as the density in a habitat increases. This assumption is justified because t is related to the resistance of the established residents. The resistance from each evidently increases. Harris (1964) shows the center of a territory of a lizard to be more vigorously defended than the edge. Thus, the per individual resistance always increases with density. Therefore, the overall resistance and t must increase with density. We shall assume that t does not vary with habitat, since it is unlikely that the territorial resistance is dependent on habitat, except as density varies.

Thus, Equation (16) may be rewritten:

$$T_i = S_i(1 - t(d_i)). \tag{17}$$

The function of T_i, like S_i, always decreases with density, since S_i always decreases and $t(d_i)$ always increases with density. Also, T_i is always less than or equal to S_i. The actual suitability of the habitat remains S_i; T_i is just the apparent suitability from the point of view of the newcomer oganism. The T_i in (17) are defined such that an ideally adapted individual will always go to the habitat where T is highest, assuming as before that all (unsettled) individuals are alike in their adaptation to the habitats. This leads, as with the S_i, to an equilibrium condition where the T_i are equal in all occupied habitats. The resulting equations, with (14), completely define a set of habitat densities and a new distribution. This distribution may be called the *ideal despotic* distribution. Like the ideal free distribution, the ideal despotic distribution is a useful basis for discussion, but because of the underlying assumptions can only approximate any real situation.

Interpretation. Some remarks about the T_i in (17) may make them easier to understand. If within an area some habitats are better than others, then the territorial restrictions may just restrict newly settling individuals to the less good habitats. Then T_i represents the average success expected in the area for individuals in the less good areas. However, in a uniform habitat where no less favorable regions exist, then the territories of all occupants will be, on the average, about equally suitable. In this case, $t(d_i)$ represents only an entering risk to unsettled individuals. If they can successfully settle in the habitats, they may expect to be as successful as the habitat average. But the act of settling may involve a serious risk of physical harm from established individuals. There may be a high probability of failure with accompanying mortality. Thus, the T_i may not be a measure of the success of some members of the habitat, but a measure of the average success of a hypothetical group of individuals which tried to enter a habitat until either successful or dead.

These two territorial hypotheses exemplify the distinction between ultimate and proximate determinants of behavior. These are defined as follows: Ultimate determinants of behavior are the environmental factors which produce the natural selection that leads to the inherited basis for the behavior. Proximate determinants are the stimuli that prompt the behavior. The behavior that we are considering is the habitat selection of individuals. The first hypothesis is that territorial behavior is *only* a proximate factor (*re* distribution) providing information about density. The second hypothesis is that territorial behavior is an ultimate factor which, by directly influencing the survival of past individuals which selected certain habitats for residence, has actually caused the population-genetic basis for the habitat selection to change in some way. And since the distribution is determined by the habitat selection, the territorial behavior is an ultimate cause of this phenomenon as well. However, in the second hypothesis territorial behavior is probably also a proximate factor. It could easily provide information about itself and provide a stimulus to which unsettled individuals could respond. This is quite likely because territorial behavior generally involves display and vocal announcement as a substantial part of the "defense" of boundaries.

The Spacing Hypothesis

Lack (1964) and Johnston (1961) have supported a third territorial hypothesis, that the territorial behavior has been evolved only to space individuals within a habitat. This means that the density in a habitat is determined independently of the behavior, but that given a certain density, the individuals separate as much as possible and have non-overlapping home ranges. For example, suppose a population of a territorial species can achieve distribution without the territorial behavior either cuing density variation or restricting habitat occupancy. Then the role of the territorial behavior may be purely behavioral, isolating mating adults to strengthen pair bonds (Lack, 1954). Or the behavior could prevent the spread of disease by a quarantine effect. Or it could maximize the consuming efficiency of the birds by lowering their search time and energy. Whatever its role, if the territorial behavior only spaces individuals, it has nothing to do with whether or not an individual will settle in one habitat or another. It only affects the movements of the individuals *within* the habitat. The effect of spacing is to keep the *instantaneous* distribution of individuals within the habitat fairly uniform. It does not alter the number of occupants of the habitats nor does it affect the average number of individuals using any portion of the area within the habitat (the average being taken over some time period). Thus, the average density over any piece of land (i.e., any habitat) is not influenced by territorial behavior. In this case, territorial behavior only spaces individuals. It has no effect on the habitat distribution, either ultimately or proximately.

The reader should be aware of the hierarchical arrangement of these hypotheses. Where Huxley's model holds, all other hypotheses hold as well. Where Kluyver and Tinbergen's hypothesis holds, Lack's also holds. Lack's hypothesis essentially excludes the others.

Dominance behavior, like territorial behavior, has cooperative and despotic possibilities. For example, sexual dominance probably has no effect on the relative fitnesses of the sexes. Dominance in feeding flocks, however, could easily result in some individuals (subdominants) having lower fitness than would be the case without the behavior. Unlike territorial behavior, however, dominance behavior cannot easily offer the advantage of cues or of spacing suggested above for territorial behavior. There is no emphasis on conspicuousness and little indication of behavioral isolation. Thus, dominance behavior can influence habitat distribution only when it is despotic, when it results in a significant *t* function and produces low fitness individuals in one habitat that disperse to another habitat. Thus, the ideal despotic model above is well defined not only for territorial despotism but for dominance despotism as well.

General Application of the Theoretical Development

Approach. The next matter is to consider the problem of identifying the actual role of the behavior relative to habitat distribution in a given species. We have defined two theoretical distributions under certain assumptions and three different territory hypotheses. We may, as a first approximation, assume that if territorial behavior has one or both of the hypothesized roles in a given species, the actual habitat distribution will be similar to the theoretical distribution described for that hypothesis. Thus, if we can show that the actual species distribution is approximately ideal-despotic, this may be taken as evidence in the support of the hypothesis that a role of territorial (or dominance) behavior is to limit density. Or if we can show that the distribution is consistently ideal-free in spite of variation in population size, and can find no other evident cues for density and/or population size variation being used by the species, then this may be taken as evidence in support of the hypothesis that the territorial behavior is used for density assessment. If we can show that the distribution is not consistently ideal-free in the face of variation in population size, nor ideal-despotic, then this may be taken as evidence for a lack of any density assessment or density-limiting mechanism, and spacing remains the most tenable of the hypotheses offered. (Other hypotheses may exist but we shall not pursue them here.) The approach to the problem rests on identifying the distribution as (approximately) ideal-free, ideal-despotic, or neither.

Density and habitat suitability. If a population has an ideal-free distribution, then by definition the suitability of all inhabitated habitats is equal. We used this to show that the ideal-free assumptions determine a distribution. However, the densities in the different habitats are not necessarily equal (see Figure 31). Thus, *if a species has a nearly ideal-free distribution, habitats with different densities will show similar success rates.*

If a population has an ideal-despotic distribution, then the suitability of all habitats is not necessarily equal. Only the apparent suitabilities (the T_i) will be everywhere equal, assuming ideally adapted individuals. In this case, it will now be shown that, for two occupied habitats, p and q, if $n_p > n_q$ then $S_p > S_q$: The suitability is higher in habitats of higher density. From the equilibrium condition, for occupied habitats q and p,

$$T_q = T_p. \tag{18}$$

Since, by (17) $T_q = S_q(1 - t(n_q))$, $T_p = S_p(1 - t(n_p))$, then substituting in (18) we obtain

$$S_p(1 - t(n_p)) = S_q(1 - t(n_q)) \text{ and } \frac{S_p}{S_q} = \frac{1 - t(n_q)}{1 - t(n_p)}. \tag{19}$$

If $n_p > n_q$, then $t(n_p) > t(n_q)$, since the t function always increases with density. Then $(1 - t(n_q)) > (1 - t(n_p))$, $(1 - t(n_q))/(1 - t(n_p)) > 1$, and therefore, by (19), $S_p/S_q > 1$.

From this it follows that $S_p > S_q$ as asserted. Thus, *if a species has an ideal despotic distribution, then the success rate in habitats with higher densities of residents will be higher.*

The above assumes that the *t* function does not depend on habitat and always increases with density. The importance of the conclusion and the extent to which it is based on this assumption suggest that we examine the assumption more closely. In fact, the *t* function depends on a balance of three factors: the amount of area that a newcomer must take away, the number of individuals that the new territory is taken away from, and the amount of resistance per territory holder. The first two of these factors balance out. When there are many individuals to contest a claim, the density is higher and the size of a standard territory, smaller. Roughly speaking, it must be as easy for an invader of a 200-acre habitat to take 1 acre away from 200 individuals as it would be to take 2 acres away from 100 individuals (same habitat). So the important element is the resistance factor per bird. At high densities, territories are smaller. It seems quite likely (if not altogether certain) that individuals fight better or are more dominant when closer to the centers of their respective territories. (Harris, 1964, shows this to be true for a lizard.) In smaller territories, the area in the territory is closer to this center and should therefore be better defended. So at high densities, *other things being equal,* resistance per bird should be higher (more effective) at high densities.

Now, one of the things that is supposed to be equal is the loss of suitability with an increase in density, as well as suitability itself. The latter is no problem as the analysis is a cross-habitat comparison. We may always start by considering *equal* suitabilities in different habitats with different densities. Then only density is a factor. If the *t* function is higher in the high density habitat, the density there will decline while it rises in the other, low density habitat. The extent of this shift may be reduced by changes in the territorial resistance due to now existing differences in suitability, but this can only take place secondarily, and the conclusion of the theory holds.

The loss of T suitability with an increase in density can confuse things if it is different in the two habitats. In fact, Rosenzweig has offered a counterexample wherein territory size (density) effects are overridden by density-dependence effects. One might encounter such situations in habitats where some totally exploitable resource, such as nesting holes, is causing the density-dependence. Then any arrangement of densities and suitabilities may occur and another way of showing despotism must be found.

It is possible to show that this relationship between density and suitability exists, using an approach that assumes territories have unequal suitability. Assume that territory size is inversely related to the number of birds against which it is defended (Huxley, 1934). The better a territory is, the more territories there are that are less good, and there are more organisms that might gain in fitness trying to take it away. Since the territory must be defended against more birds, it must be smaller, Hence, more suitable territories are smaller, and density, which is inversely related to territory size (Zimmerman, ms.) is seen to be higher in high-suitability habitats.

Figure 33 shows examples of the S_i and T_i curves for $N = 2$ and a sample population distributed freely (filled circles) and territorially (open circles). Note that the suitability in the ideal free distribution is equal in both habitats ($S_1[\bullet] = S_2[\circ]$). In the territorial distribution, both the suitability and the density are higher in habitat 1 than in habitat 2; ($S_1[*] > S_2[*]$, $n_1 > n_2$).[3]

[3] We should note the discussion of Gibb (1961) in which attention is drawn to the relationship between the role of territorial behavior and differences in habitat suitability. Gibb's remarks are not formally developed and do not distinguish the density assessment and density-limiting hypotheses. But they do foreshadow several of the ideas presented above.

These conclusions suggest that the role of territorial behavior in the habitat distribution of a territorial species can sometimes be ascertained as follows: If the high density habitats show consistently higher success rates, and if no density limiting mechanism other than territorial behavior is evident, then the role of the territorial behavior is evidently to limit density. If suitabilities in all habitats are equal, even though densities are not, if the distribution changes with changing population size, and if no alternative density or population size cue is apparent, then the role of the territorial behavior is evidently to serve as a density assessment mechanism. If neither of the above two criteria is met, then the role of territorial behavior seems to be only to space individuals. Underlying this approach is the assumption that the organisms are approximately ideally adapted. This assumption may well fail and this possibility should always be considered. There is also the possibility, ever present, of hypotheses different from those considered here. Brown (1969a, b) has developed a line of reasoning similar to the above. With somewhat less formal logic, Brown did not arrive at the suitability-density correlations, nor did he recognize nonaggressive, epideictic territorial behavior. He did, however, consider the territorial influence on nonbreeding, as this varies with population size. Brown's reasoning thus complements the above (and vice versa), as he has related population size-dependent effects of territorial behavior on reproduction, while I have related population size-dependent effects of territorial behavior on distribution.

Uncertainties in application. There are a number of uncertainties involved in these conclusions which should be given careful consideration. Some of this uncertainty is inherent in the theory, which describes the relationships of expected values of population means. Any given realization of the theory (e.g., the observed habitat suitabilities in a given year) is expected to deviate from the average values, even if the assumptions of the theory are met. This deviation will alter the succeeding year's distribution somewhat (by accidental selection) so some fluctuation in the distribution may also be expected.

Another source of uncertainty lies in inherent failure of the assumptions, particularly the ideality assumption. For example, the sensory reception of the organism cannot be perfect so the organism can be expected to misread whatever environmental cues it uses to assess the suitability of a habitat. Also the correlation between these cues and suitability is probably never perfect. Finally, the species may reasonably be expected to be always evolving towards the ideal state without ever achieving it. These failures of the ideal assumption will lead to errors in the individual judgments of habitat suitability.

Hopefully the uncertainties in our predictions will be generally independent in different years, or even in different regions in the same year. Probably most of the error-causing factors are rather local in effect. A major exception is weather, but even this factor changes from year to year. Thus, in most cases, sampling over several years or perhaps over widely separated regions in the same year should provide a way of estimating or controlling these errors. If the errors are not independent even over years, then the species may be considered to be evolving and changes over time should be detected. Thus, results which are consistently obtained over a number of years may reasonably be considered free from these errors.

Measurement of suitability. The problem of measuring suitability remains. The suitability of a habitat is a reflection of the average genetic contribution of resident adults to the next generation and must be closely related to the average lifetime production of reproducing offspring in the habitat (i.e., $R = e^{rt} = \Sigma_x l_x b_x$). Therefore, it must depend on several components, including reproductive rate and survival of adults and immatures. Since territoriality is normally associated with breeding behavior and habitat distribution during the breeding season, we shall usually associate suitability with such things as nesting success, feeding rates, and fertility.

Specific Applications of the Theory

This theory can be used qualitatively or quantitatively. We can, for example, predict from the expected sigmoid nature of the suitability curves that in the ideal free distribution the changes in density in low-density habitats

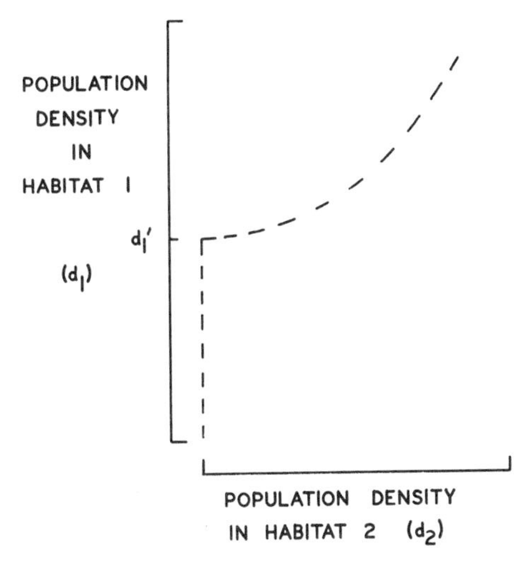

FIGURE 34. Theoretically predicted plot of densities in the first two habitats with variation in population size. Densities in habitat 1, the "best" habitat, increase while no birds occupy habitat 2, up to density d_1'. At this density, $S_1 = B_2$ and birds occupy both habitats. At first, due to a presumably sigmoid suitability-density curve, the increases in habitat 2 exceed those in 1 (see Figure 30) and the curve is flat. Later, the relative increases are similar, causing the curve to bend upward.

with changes in density in high-density habitats will follow a general pattern as in Figure 34. This pattern can be compared with data plotted from Kluyver and Tinbergen (1953, Figure 35) and from Figure 36.

We can also, as already noted, examine density-suitability correlations in order to assess the role of behavior in distribution. For example, in polygynous birds, suitability for a male is dependent on the number of females he has. Thus the sex ratio in a habitat is a partial measure of the suitability of the habitat for males. Fretwell and Calver (1970) measured sex ratios of dickcissels in different habitats and in different geographic regions with results as shown in Figure 37. The dickcissel densities were highest as points D (III) in the Mississippi River valley, and higher

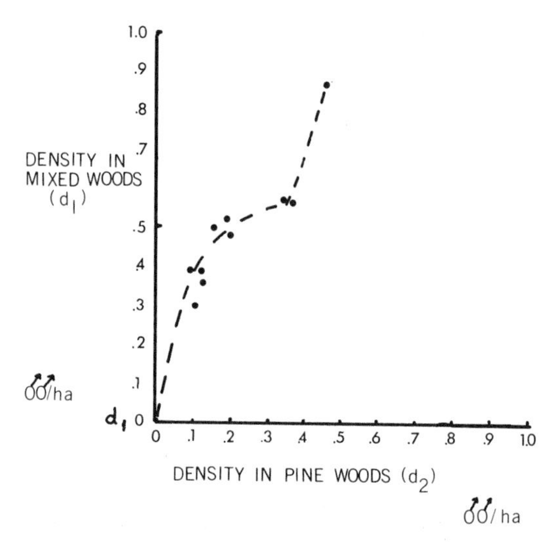

FIGURE 35. Distribution of coal tit at Hulshorst. This graph suggests that $B_1 = B_2$, but that density-dependent effects are considerably more severe in habitat 2 (pine woods). This is so because the extrapolation to $d_2 = 0$ of the dashed line is nearly $d_1' = 0$. (Compare with Figure 34.) At d_1', $S_1 = B_2$, also, at $d_1 = 0$, $B_1 = S_1$ so if $d_1' = 0$, $B_1 = S_1 = B_2$. (From Kluyver and Tinbergen, 1953.)

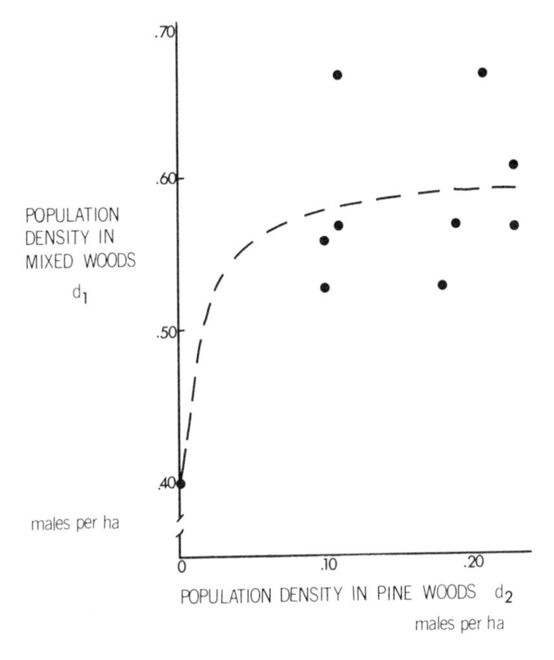

FIGURE 36. Distribution of great tit at Hulshorst. These data indicate that $B_1 > B_2$, that mixed woods are superior to pine woods in density-independent effects. There is no evidence that density-dependent effects ever become important in pine woods. (From Kluyver and Tinbergen, 1953.)

in clover fields than in pastures. The sex-ratio index (percent of males mated) was consistently higher in areas or habitats where the density was higher, supporting the density-limiting theory of Huxley (1934). Zimmerman (1971) has recently confirmed this.

I have attempted to apply the same theory to a winter population of juncos (*Junco hyemalis*) (Fretwell, 1969a). Figures 38, 39, and 40 give the results: There was a clear dominance-survival effect (Figure 38) so the fitness of some individuals was apparently reduced below the group average by dominance effects which of course are directly

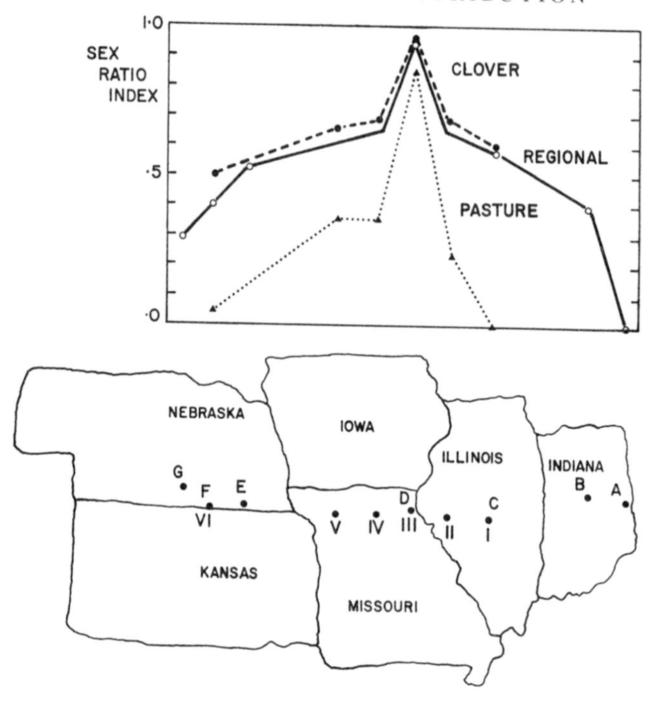

FIGURE 37. Local and regional variation in a sex-ratio index of the dickcissel. On the map are shown the study areas. Above, the estimates of overall regional indices are plotted as open circles and connected with a solid line. The estimates of regional clover field indices are plotted as solid circles and connected with a dashed line. The estimates of regional pasture sex-ratio indexes are plotted as solid triangles and connected with a dotted line. The graph of the overall regional sex-ratio index is drawn below the graph of the clover field sex ratio. (From Fretwell and Calver, 1970.)

related to density. This establishes the existence of a large-valued t function. So we may presume that it would be to the advantage of some of the subdominant juncos to disperse to habitats which are, on the average, less suitable than the one occupied at a high density of more dominant individuals. Figure 39 shows that weedy fields are occupied at high density and that forested areas are occupied at low density. So we expect the subdominant juncos to go into

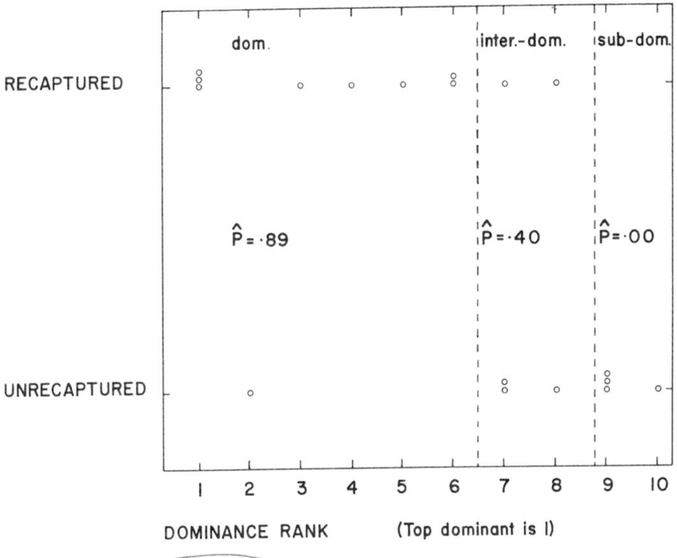

FIGURE 38. Recapturability and dominance status. Recapturability reflects survival, and dominance status may reflect order of arrival in habitat (Sabine, 1955). Higher ranks would then be late arrivals. When established bird density is higher, these ranks are higher and late arrivals suffer a greater loss of fitness. (From Fretwell, 1969a.)

the forest. In Figure 40, dominance is shown to be somewhat related to wing length (larger juncos are more dominant), so we expect the smaller juncos to go into the forest habitats. The average wing length of the forest juncos was, as given in Figure 40, 75.8 mm, while the field average was 77.8 mm, confirming this prediction. Thus, in this winter population the habitat distribution appears to be ideal-despotic, not ideal-free. In the great tit, however, it appears that poor and rich habitat birds might be equally successful, due to density adjustments (Kluyver and Tinbergen, 1953). Krebs (1971) has challenged this, noting that in low-density habitats (pine woods), several components of suitability (clutch size, survival of young) are lower. Krebs' argument is weakened by the fact that he did not make

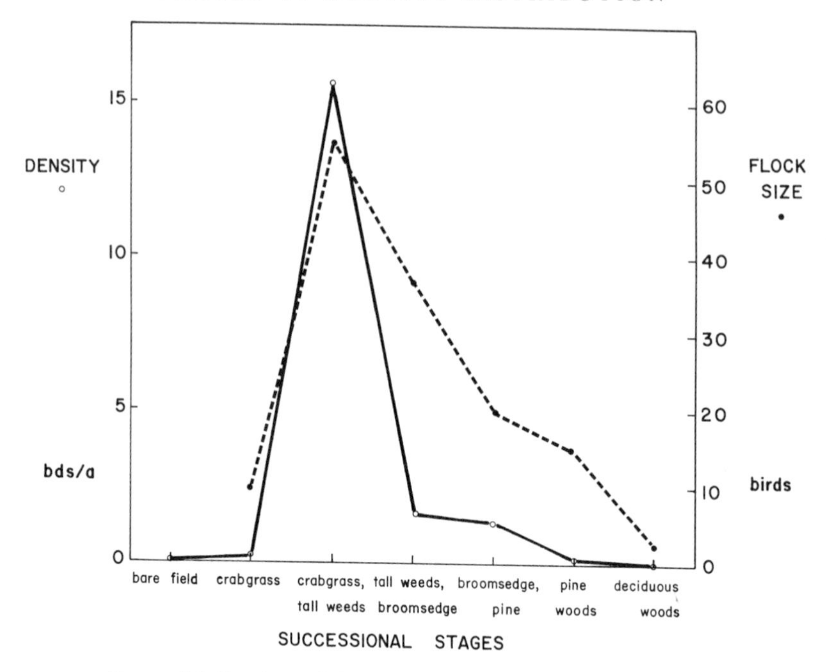

FIGURE 39. Density and flock size of junco over plant successional stages at Raleigh, North Carolina. (After Quay, 1940.)

comparisons within a year. In the ideal free distribution (equal success rates in both habitats), the low-density habitats are occupied only in years of high population, when every pair is doing less well, regardless of habitat. Thus, data from pine woods, on the average, will reflect poorer success rates than data from deciduous woods, but in a given year the success rates will be equal. When rates are high in deciduous woods, there would be no data from pine woods. Thus a comparison of average success in different habitats is generally invalid. The comparisons must be made within a year.

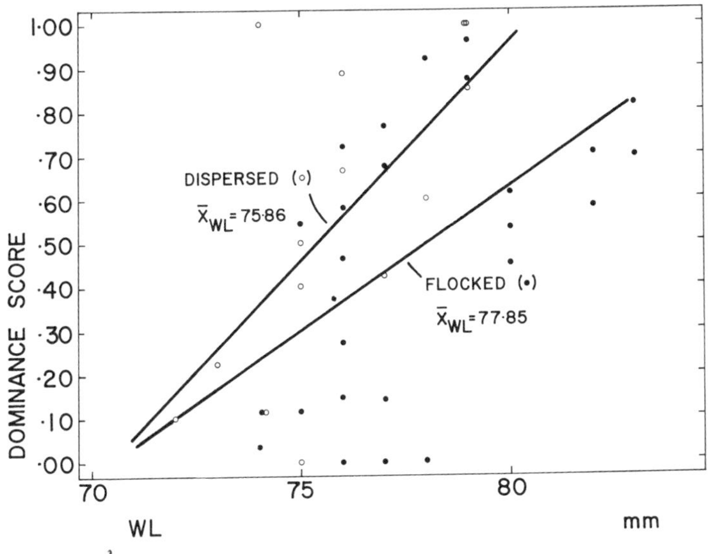

FIGURE 40. Dominance and wing length. (From Fretwell, 1969a.)

SUMMARY

Habitat distributions in dispersive species are set up via habitat selection. Given density-dependent suitability in each habitat and ideally adapted habitat selection behavior, then corresponding to every population size there is a unique set of habitat densities optimizing evolutionary success. This distribution is referred to as the ideal free distribution; it can be achieved only if every individual is free to express its habitat choice.

If suitability initially rises with density and then declines (Allee-type effects), then the optimum distribution may be a very unstable function of population size where slight increases in numbers will affect gross changes in distribution.

Dominance as territorial behavior can play one of several roles in distribution. Territorial behavior can be "epideictic," indicating density levels to permit refined habitat selection. Territorial and dominance behavior can also be despotic, forcing individuals away from the habitat of their choice and so changing the distribution from the optimum form. This behavior may also be irrelevant with regard to distribution.

If dominance or territoriality are despotic, there may well be a positive correlation across habitats between suitability and density. Otherwise, suitability should be equal in all habitats.

Breeding male dickcissels in the midwest appears to use territoriality despotically. Also, wintering juncos appear to use flock dominance to alter the distribution from the ideal form.

CHAPTER SIX

Distributions of Breeding Birds

Bird population studies are inordinantly common when one considers the relatively unimportant part birds play in ecological energy flow. This could be due simply to convenience—birds are easy to study so they *are* studied. However, it is also possible that birds are really the appropriate animal to use for the answers to some of the questions that lead us to consider population ecology at all. One tends to be anthropomorphic with birds, so this group assumes special popular status. Conversely, it is easy (for me, at least) to sympathize with birds, to feel that they have somewhat analogous problems and pressures, and that we can gain insight about ourselves from studying them. It is not surprising to note that one of the first advances in bird population ecology, the territory theory, as advanced by Howard (1920) is now being applied (or misapplied) to human behavior (Ardrey, 1967).

Whatever our ulterior motives, birds are easy to study. When fresh data must be collected to apply a theory, the "moderately" competent field biologist is more likely to have success with birds than with any other group.

The studies that follow are presented chronologically. They were conducted as an example of the general usefulness of certain theoretical ideas, but have acquired, I think, some interest of their own. Whether or not this is so, the presentation here attempts to preserve the approach used to facilitate future applications.

DISTRIBUTIONS OF BREEDING BIRDS

PART I—STUDIES OF THE FIELD SPARROW

Breeding Distributions

A bird species which is found in several habitats in the breeding season probably achieved that distribution through dispersion and habitat selection. From the preceding chapter, we know that if fitness or success in the habitats is density-dependent and if the birds' habitat selection behavior is responsive to all factors affecting potential success, then the habitat distribution of the species is determined by the fitness-density relationships and the population size. In cases where the fitness-density relationship stems from territorial or dominance behavior, it is easy to suppose properly responsive habitat selection behavior in the organisms. The aggressive behavior provides easily interpretable information; the subdominant bird can hardly help but respond appropriately to a physical attack.

The problem is not, in general, as obvious in the simple ideal-free case, where aggressive contributions to density-dependence are not of major importance. A proximate factor to which birds can respond need not exist, nor are the responses of the birds to whatever cues that are available limited to the appropriate ones.

In my first attempts to apply these ideas, I wished to examine the easy case first and to study a territorial species. A choice did not exist however; the field sparrow (*Spizella pusilla*) was the only available species convenient and easy to study, although it does not show clear signs of aggressive territorial behavior in central North Carolina.

The prediction of the ideal free distribution is equal rates of success in habitats occupied at different densities. Initial field data for the field sparrow (from Fretwell, 1970, Table 3) did not bear out this prediction. The field sparrows did occupy two habitats at different densities, but the factors affecting nest success were generally more favorable in the habitat occupied at low density.

In order to make such a test of the prediction, it is necessary to assume that the habitat selection of adults is ideal and that the measures taken of success were appropriate. Either or both of these assumptions may be invalid. The best way to find this out is to assume that they are invalid and then to predict the outcome given *this* assumption. In the first case, if the field sparrow is maladapted, so individuals overcrowded habitat 1 in Table 3, then these individuals should be selected against and should become relatively less abundant in the population. However, the censuses in succeeding years looked similar to the first year's census (Tables 3 and 4), showing no selection favoring the more successful breeders. So we need not accept the assumption that birds are maladapted. I shall now proceed by considering what could determine suitability besides nesting success.

Winter Survival

Fitness is determined by the balance between birth and death. If measurements on nesting production are not complete measures of breeding habitat suitability (by assumption) then measurements on survival must be needed for a proper estimate. There is very little mortality in the breeding season because breeding takes up such a short time. Also, in Table 3, some measures of survival in the breeding season agree with the measures of reproduction in favoring the less populous habitat. The implication then is that factors affecting survival which operate outside the breeding season are important and that the interaction of the field sparrows with these factors depends on the breeding habitat, or breeding experience, or both together.

Three incidental observations become relevant here. The first concerns Wynne-Edwards' hypothesis, regarding

Table 3. Habitat differences in breeding biology of field sparrows (for 1964 only)

Winter suitability	Null (2)		High (1)	
Density	10.1 ha	138.4 ♂♂/100 ha	5.74 ha	397.8 ♂♂/100 ha
Nest mortality	172 nd †	.029 failures/nd	333 nd	.073 failures/nd
Clutch size	10 n(ests)	3.56 eggs/n	17 n	3.43 eggs/n
Egg mortality	10 n	0.0 failing eggs/n	17 n	.7 failing eggs/n
Desertions *	172 nd	.006 desertions/nd	333 nd	.015 desertions/nd

* Suggested female mortality.
† nd is nest-day.

Table 4. Habitat differences in breeding biology of field sparrows (for 1965–1966)

Winter suitability	Null (2)		Low (3)		High (1.4)	
1965						
Density	14.2 ha	170.5 ♂♂/100 ha			5.74 ha	397.8 ♂♂/100 ha
Nest mortality	109 nd	.028 failures/nd			160 nd	.056 failures/nd
Clutch size	8 n	3.88 eggs/n			11 n	3.35 eggs/n
Egg mortality	5 n	.2 failing eggs/n			7 n	.28 failing eggs/n
(May only)	5 n	9.9 gm/yng			5 n	9.6 gm/yng
1966						
Density	16.2 ha	74.1 ♂♂/100 ha	10.1 ha	217.5 ♂♂/100 ha	4.05 ha	321.2 ♂♂/100 ha
Nest mortality	15 n	.67 failures/n	17 n	.70 failures/n	10 n	.80 failures/n
Clutch size *	8 n	3.86 eggs/n	12 n	3.92 eggs/n	8 n	3.12 eggs/n
(May–June)						
Brood size †	5 n	3.5 yng/n	5 n	3.2 yng/n	4 n	3.0 yng/n
Broods/pair	9 pairs	.89 broods/pair	19.5 pairs	.54 broods/pair	12.5 pairs	.44 broods/pair

* First nestings.
† Successful nestings.

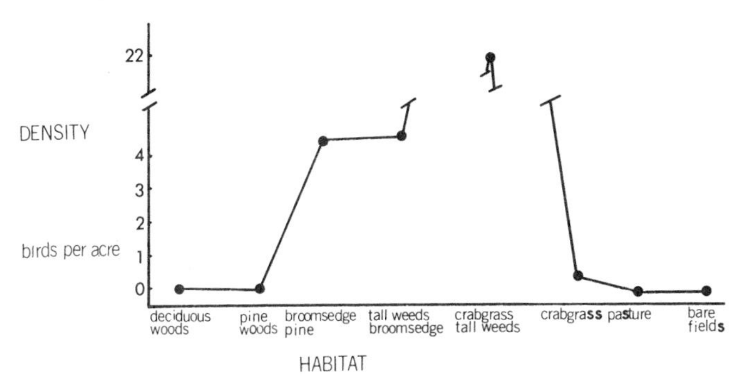

FIGURE 41. Densities of field sparrows in some winter habitats in North Carolina. Field sparrows are absent from forest, pasture, and bare field habitats. Although densities appear to be highest in crabgrass and tall weeds, such habitats are so rare that only a modest proportion of the species winters there (personal observation; see also Quay, 1947, p. 383). The bulk of the species is limited to broomsedge habitats. (From Quay, 1940.)

the role of social behavior in animal dispersion (1962). Whether effected through group selection or not, Wynne-Edwards' ideas stress the possibility of social "ecotypes" in populations. From his arguments, it becomes important in terms of fitness for an animal to be socially competent, to prepare itself, for example, for dominance contests.

The second observation comes from the good winter census data of Quay (1947). The overcrowded habitat in Table 3 was a pine-broomsedge habitat (Oosting, 1942). This habitat is very similar in appearance throughout the year. From Quay's data (Figure 41), it sustains most of the populations of field sparrows in the winter.

The third observation arose from studies which I made on an aberrant Eastern colony of dickcissels (Fretwell, 1967). These birds were well outside the normal breeding range of the species and so were expected to do poorly. There were only three pairs and as many nests, but there was no indication of the anticipated difficulties in nesting. The birds chose the least utilized habitat available and fledged two of three nests, or seven young. Feeding rates were high. These birds did much better, in fact, than any other recorded dickcissels from the normal nesting range, where cowbirds and predation severely limit production (Zimmerman, ms.). I concluded that an increase in total dickcissel population with additional birds settling in the East might not result in a reduced average birth rate, and that there was little evidence for population size (density) dependent success in the breeding season. If so, then the population must be limited by the winter season.

These three observations led to the development of the following hypothesis for the field sparrow (Fretwell, 1968):

1. Field sparrows are intensely limited by *winter* resources, specifically broomsedge (*Andropogon virginicus*) seeds.
2. Field sparrows regulate the utilization of these seeds with social (dominance) interactions as argued by Wynne-Edwards (1962).
3. Field sparrows breed in the winter habitat type as a means of enhancing winter social status.

Predictions from this hypothesis are:

a. Individuals which breed in different habitats winter together.

b. Individuals which breed in the winter-type habitat would, in the winter, have more fat.

c. Fat birds would survive better and would be recaptured more frequently.

d. Annual survival will be higher for birds which breed in the wintering habitat.

These predictions were all verified.

a. Of 23 males breeding in the wintering habitat, 14 (60 percent) were banded in the winter habitat the previous

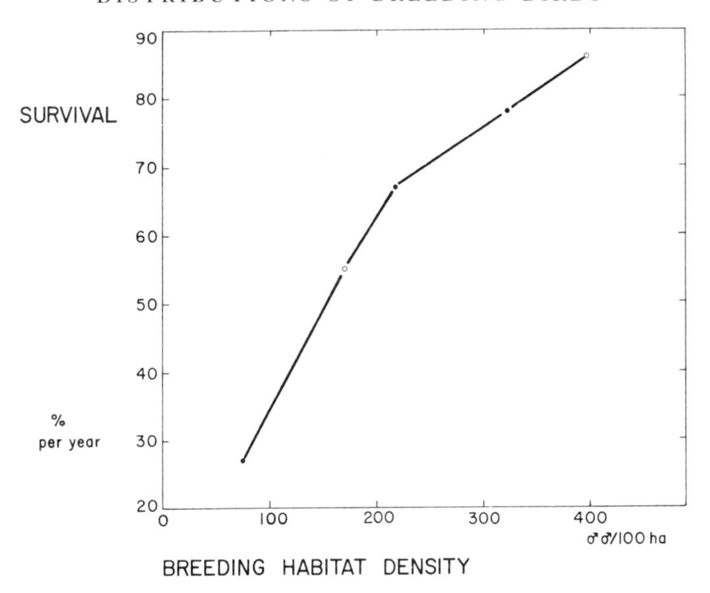

FIGURE 42. Survival and breeding habitat density. Open circles represent estimates obtained by comparing the proportion of breeding residents which were winter residents. Closed circles represent percent return to territory. All estimates are confined to males. (From Fretwell, 1968.)

winter. Of 24 males breeding in nearby nonwintering habitat, 13 (54 percent) were also banded the previous winter in the winter-type habitat.

b. Out of 12 males which wintered and bred in the winter type habitat, 9 (75 percent) were higher in fat class than the average of the group in which they were caught. Of 7 recaptured males which bred outside the winter habitat, only 1 (14 percent) was higher in fat class than the average of the group in which it was caught.

c. About 67 percent (21 of 33) high fat birds were recaptured after a month of winter, while only 32 percent (11 of 34) low fat birds were recaptured.

d. In Figure 42, breeding density, which is higher in winter-type habitats (Table 4), is plotted versus annual survival. Breeding residents which were winter residents are presumed to be the only adults aged two or more years. The proportion of nonwinter residents estimates the rate at which one-year-old birds (dispersive and therefore not resident in winter) enter the breeding population. This, of course, is equivalent (if the population is stable, as it nearly was) to the rate at which older birds die from the breeding population. One minus this rate then yields an estimate of annual survival.

Table 4 summarizes additional breeding data for the field sparrow, confirming the original suggestion that total breeding success was higher in the wintering-type habitat.

Some of the predictions of the hypothesis were untestable in the field sparrow, which has obscure dominance relationships in winter as well as in the breeding season. So the role of dominance in wintering flocked sparrows was studied in the junco (*Junco hyemalis*). We have predicted that subdominants will not survive as well as dominants. This prediction is verified in Figure 38. Also, more dominant juncos were fatter (Fretwell, 1969a).

Overall Fitness

Granted the hypothesis formulated above, it would be clear why some field sparrows choose to breed in the winter-type habitat (not that it is certain: this conclusion depends on the hypothesis, which is never proved). By breeding in the winter-type habitat a field sparrow gains a winter survival advantage. But some field sparrows breed elsewhere. This, from the theory of Chapter 5, indicates that some of the birds are responding to density-dependent factors in the breeding habitats, avoiding the best habitats as these become overcrowded. We can now test again the ideal free prediction of equal suitability in the habitats occupied at differing density. This can be accomplished by assuming that fledged young per lifetime is a good measure of fitness. We can estimate this measure and do so here to provide an example.

Using only 1966 data, we note from Table 4 that the number of broods per pair at the end of the first half of the nesting season is .89 in a habitat of no winter suitability (WS_0), .54 in a habitat of medium winter suitability (WS_L), and .44 in good wintering habitat (WS_H). Doubling these values and multiplying times the brood size yields total per pair per year production in each habitat. These figures are 6.23 (WS_0), 3.45 (WS_L), and 2.64 (WS_H). The number of offspring per adult per lifetime depends on the expected number of breeding seasons per adult, which depends on the rate of survival. Counting from the first breeding season, the probability of only one such season is $(1 - S)$ where S is the rate of survival. The probability of two such seasons is the probability of the first and the second, but not the third. Such a bird would have to survive the second year, but die the third. The probability of the first such event is S, of the second is $(1 - S)$, of both, $S(1 - S)$. In general, a fraction $S^{n-1}(1 - S)$ of the birds enjoy n breeding seasons. The average number of breeding seasons is therefore

$$\sum_{n=1}^{\infty} ns^{n-1}(1 - S) = \frac{1}{1 - S}.$$

Thus WS_0 birds average only 1.4 breeding seasons ($S = .287$), WS_L birds ($S = .69$), 3.205, and WS_H birds ($S = .77$) 4.329. Then the total young produced per lifetime per pair is for WS_0, $1.4 \times 6.23 = 8.74$; for WS_L, $3.205 \times 3.45 = 11.06$; for WS_H, $2.64 \times 4.33 = 11.43$. Thus birds breeding in the winter-type habitat, although producing only half as many young per season, might produce about 2.7 more young per lifetime. This difference may be real but is certainly not statistically significant. If survival in the nonwintering habitat were as high as .42, the difference would be eliminated. This represents an error of estimation of only .13 from the estimated value, .286. With only 7 birds in this one group, the standard error of the mean estimate given is about .18.

The similarity of the three figures lends some confidence to the theory that the distribution over the studied habitats is regulated to equalize expected success rates. However, since the mortality of the young after fledging is not considered, a major component of fitness or suitability remains unknown. Estimation of this component could easily reverse all conclusions.

Density-Dependence in the Breeding Season

Accepting young produced per lifetime as a measure of fitness, we would like to fulfill the theory of the ideal free distribution by drawing the fitness-density curves for the several habitats. We can accomplish this if we assume that, as far as breeding goes, all the habitats studied above are equally good for a given density of birds. That is, we can assume that the differences in fitness between habitats are caused only by differences in expected length of life. We also assume that this expected length of life does not vary with density and is fixed for each habitat. Then we can estimate the expected lifetime production of offspring as a function of density for each of the three habitats. This is accomplished simply by taking the observed annual production data and associating these figures with the density in the habitat where they were obtained. For example, from Table 4, in habitat WS_0, annual production was 6.23 young and density was 74.1. Therefore, production, P, at 74.1 males in habitat WS_L is 3.205 seasons $\times$ 6.23 young/season $=$ 19.97 young. The analysis is plotted in Figure 43.

This analysis is dissatisfying in that we have had to as-

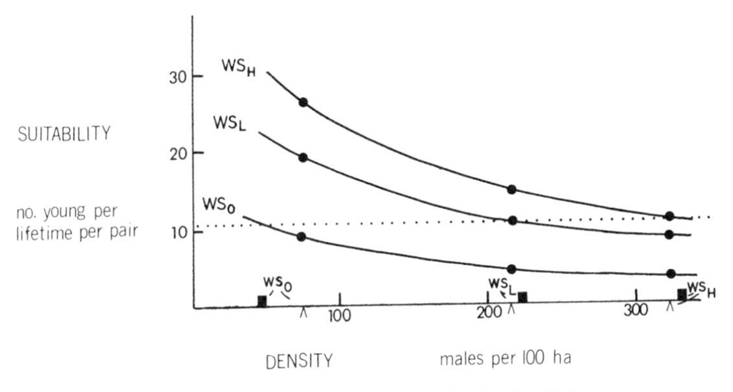

FIGURE 43. Density versus habitat suitability in the field sparrow. The solid curves are g versus N curves, computed as described in the text. The lowest curve is habitat WS_0, the highest, habitat WS_H. The dashed line determines the ideal free distribution given a fledgling survival rate of 18 percent and a lifetime per pair production of two breeding adults (i.e., population equilibrium). The blocks on the density axis mark the density coordinates for this distribution, for the three habitats. For example, the leftmost block is the density in habitat WS_0 that will produce eleven offspring, or two breeding adults. The actual densities are shown by the arrows; the data are not good enough to permit evaluation of the differences.

sume that habitats are equivalent in the breeding season as far as nesting success is concerned. The demonstration of density-dependence of fitness in the breeding season however, depends on this assumption. It is unfortunately possible that the different rates of production in the several habitats are due to differences between the habitats. In order to assure ourselves that within habitat nesting success is density-dependent, the following refined analysis was applied to the data collected in field #1 in 1964, described in Fretwell (1970). For each 24-hour period of the study, the total number of known, active nests was separately tabulated. This number was used to index the nest density in the habitat during that daily period. For each of these periods, the known number of nests present each contributed one nest-day (Mayfield, 1961). In each period there were a certain number of nest failures, from zero up to the total number of known nests. The nest-days and the failures were summed over all periods with the same total number of nest-days, that is, the same density index. For example, suppose there were ten 24-hour periods when only one active nest was present. Suppose further that during just one of these periods did the single known nest suffer mortality. Then the summed totals for periods of density index one was $1 \times 10 = 10$ nest-days, and 1.0 failure. Using Mayfield's method, the daily mortality rate for periods of this density index is $1/10 = .1$ failures/nest-day. If there were also ten 24-hour periods when four nests were known to be active, then the summed total of nest-days for periods of index four would be $4 \times 10 = 40$ nest-days. If four nests failed during these ten periods, then the daily mortality rate estimate would be 4/40 or .1 failures/nest-day, for periods of density index four.

From zero to seven nests were known to be present daily in the field #1 in 1964, so periods were indexed one to seven. Presumably these indices well represent the actual variation in nest density in the field. There were several occasions when nests were not checked every 24 hours and mortality occurred. The mortality in these cases was arbitrarily assumed to have occurred in the middle of the lapse. When more than one predation occurred, they were assumed to be distributed evenly over the lapse. This latter procedure assumed an inverse density-dependent mortality with lower rates at high densities. Therefore, insofar as it is used, the results will be biased towards a negative relationship between density and nest mortality (opposite to the prediction, therefore conservative). The problem occurred only during periods of high index, which were rather unusual. The data were retained in spite of the bias in order to have a larger sample of periods with this high index.

The final step of the analysis was a correlation between the nest density indices and the estimated daily mortality rates for each index. Density-dependent nest mortality would appear in a positive correlation between the variables. The analysis was tested for statistical significance using a weighted regression procedure (Steel and Torrie, 1960).

Results and Discussion

The results are presented in Figure 44, with the regression line drawn through the points. There is a clear positive relationship between daily nest mortality estimates and the density indices. This relationship is statistically significant

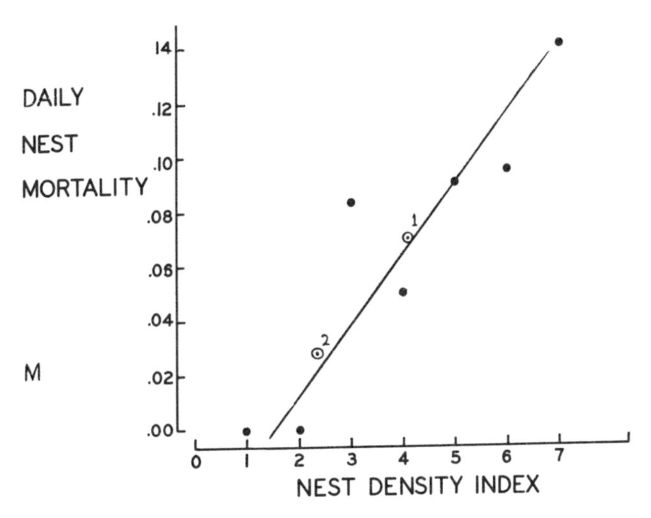

FIGURE 44. Daily nest mortality versus density. The points (uncircled) are data from habitat 1 (Table 1) and are from 40 nests found there (25 mortalities). The within-habitat nest density varied from 1 to 7 in habitat 1. The daily mortality rate, M, per nest was estimated by the formula: $M =$ (number of mortalities on days when density is d_1)/(days when density is d_1) · (d_1). The circled points are averages from habitats 1 and 2 (Tables 3, 4). Their proximity to the significant regression line suggests that little or no difference between habitats in predation exists that cannot be accounted for by differences in density.

($t = 4.84$, 5 d.f.). This is true despite the bias toward a negative correlation mentioned above.

One of the difficulties with the present analysis is the fact that the different indices occurred on different days. Actually, the nest density varied in a consistent pattern over the breeding season. There were two peaks of frequency of high index periods. Thus, if there was a temporal pattern of nest mortality that was by chance correlated with the temporal pattern of nest density, the results shown in Figure 44 might not represent a causal relationship. However, the analysis of these data in Table 4 compared habitats occupied by field sparrows at different densities; that analysis also found a positive correlation between density and nest mortality, but little or no seasonal bias existed in these data. These data were vulnerable to any habitat bias that might exist, but unless both habitat *and* temporal biases exist, the combined analyses indicate that a high nest density results in a high nest mortality, in the field sparrows at least.

Presumably, the phenomenon demonstrated in Figure 44 is similar to that described by Tinbergen (1960). In that paper he postulated that predators develop searching images of the prey that allow the predator to scan large areas ignoring all other irrelevant stimuli. The images are reinforced by conditioning after a successful catch so the predator becomes more efficient at spotting a recently caught prey type. The more efficient predator then raises the mortality rate for this particular prey. The probability that a predator will catch the first of a particular prey type is dependent on the density of the prey type. If one type of prey is very abundant, the predator is likely to encounter one of them. Then, by conditioning, the predator's image is reinforced, and the mortality rate of the prey increases. Thus, at higher densities of the prey, the mortality rate of the prey is apt to be higher (see Holling, 1966).

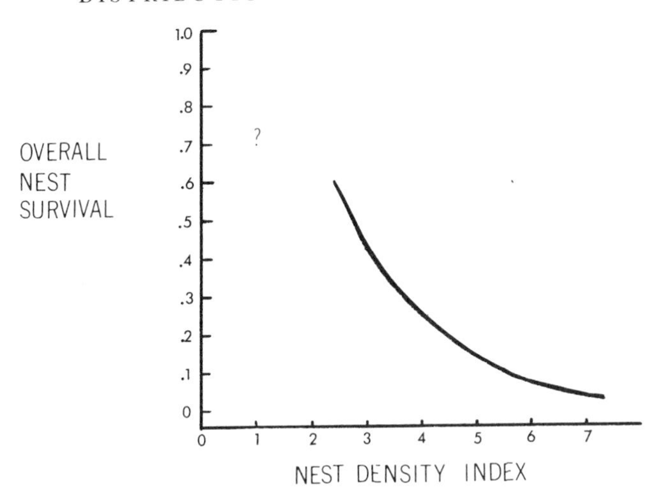

FIGURE 45. Overall nest survival as a function of density.

The predators of field sparrow nests surely include blue jays (*Cyanocitta cristata*), a rodent or insectivore, and an unidentified black snake, all of which have been seen at the nests. Of course, other predators may be involved as well.

We began this analysis in order to test the assumption that density-dependence exists within habitats as well as between habitats. We have also assumed, however, that habitats are equal with regard to breeding success. This assumption is tested in Figure 44, with the lower circled point. This point represents the habitat averages for habitat WS_0, 1964 (number of nests present per 5.74 ha versus average daily nest mortality). This point is quite close to the regression line, indicating that any differences in nest mortality between these habitats that are not related to density are too small to be detected by the present data.

In Figure 45, I have translated Figure 44 into percent nests succeeding by plotting $(1 - y)^{21}$. The daily rate of survival is $(1 - y)$ and there are about 21 days in the nesting cycle. The *y* values were taken from the linear regression line.

When these analyses are interpreted in the framework of the preceding theory, we obtain the following description of the regulation of habitat distribution in breeding field sparrows:

Different breeding habitats have different basic suitabilities depending on their similarity to the winter habitat of the species. Within a habitat, breeding success is density-dependent, mainly because of density-dependent nest predation. This effectively limits the density in the habitats with the highest basic suitability (i.e., the habitats similar to the winter grounds). The habitat selection of individuals is to some degree properly responsive to this limitation. Thus, there is dispersion to other habitats with lower basic suitability. These other habitats should be settled to an extent dependent on the population size of field sparrows in the spring, according to the curves drawn in Figure 43. Although territorial effects are not clearly demonstrated in this dispersion, they may play some role.

In this interpretation, no distinction is made between dispersion and migration, except perhaps one of distance. In the spring, field sparrows evidently "disperse" thousands of miles to find habitats which are unsuitable in winter (and unoccupied). This dispersion (migration) is typically northward, as northern habitats have a very low winter suitability.

Interspecies Relations

The habitats in which the studies of field sparrows were conducted were occupied by a number of other species, none of which was nearly as common as the field sparrow. I nonetheless collected a small amount of nest-success data on some of these species.

The data are summarized in Table 5. Note first that most

TABLE 5. Relationship between nesting success of uncommon open nesting species, and the density of the abundant field sparrow

Habitat 1 (high density of field sparrows)				Habitat 2 (low density of field sparrows)			
Species	# nests	# success	%	Species	# nests	# success	%
Yellowthroat	10	3	30	Yellowthroat	3	2	67
Prairie warbler	5	0	00	Dickcissel	2	2	100
Indigo bunting	2	0	00	Indigo bunting	2	1	50
Blue grosbeak	3	0	00	Blue grosbeak	3	2	67
Towhee	8	1	12	Towhee	1	1	100
Cardinal	5	2	40	Meadowlark	2	1	50
Total	33	6	18	Total	13	9	69

Unpublished data.

of the other species nesting with the field sparrow in the old field habitats were quite sparse, considered species by species. In most cases, each species never had more than one nest in a field at a time. The yellowthroat and towhee are exceptions to this rule, but even these species were less than one-fourth as abundant as the field sparrow. Yet the nest predation rate for all these other species, rare as they were, was apparently not significantly different from the rate for the field sparrow. If nest mortality is density-dependent within species, as suggested by Figure 44, then the nests of these sparser species ought to survive at a higher rate than the nests of the dense field sparrow. Apparently this is not the case. We therefore reject the hypothesis that the trend found for the field sparrow applies independently to other open-nesting species exposed to the same predators. Now note that in habitat 2, where the field sparrow was less dense, the nests of the other open-nesting species generally survived better than they did in habitats 1 or 4, where the field sparrow is more abundant. The densities of these species in habitat 2 were generally lower than in habitats 1 or 4. But densities in both habitats were so low for these species that the differences in nest survival between habitats can hardly be accounted for by the differences in density, at least according to Figure 44.

It is possible that the density of predators was higher in habitats 1 and 4, explaining the high mortality there. This alternative has been tested above (Figure 44). If habitat 1 really has more predators than habitat 2, then the nest mortality in habitat 2 should be lower than the value in habitat 1 at the same density. The value in habitat 1 at a low density can be estimated from the regression line in Figure 44. This comparison is made on the figure, and as can be seen, the value in habitat 2 (lower circled point) is very close to the value on the regression line at the same density. The level of predation, on the field sparrow's nests at least, seems to be similar in the two habitats.

The nests of different species of birds are often rather similar, especially when seen from a distance. Differences in location are often small and differences in construction even less evident. Identification of different nests as to species is in many cases possible only by close inspection. Suppose the searching image hypothesis of Tinbergen discussed above is a valid explanation of Figure 44. Then it seems possible that the searching image developed for one species' nests may aid the predator in finding nests of other species. Then the density of nests of one species would affect the mortality not only of its own nests but of nests of other species as well. In fact, the nest mortality of each open nest would be dependent on the density of all open nests, irrespective of species. This would account for the observations in Table 5. According to this hypothesis the rare species have nest mortalities similar to the abundant field sparrow because they are as much affected by the field sparrow's nest density as the field sparrow itself. The nest mortalities of these rare species is lower in habitat 2 than in 1 or 4 because the density of the field sparrow is lower there *and* because there is no other abundant, open-nesting species in this habitat. The total number of open nests is lower in habitat 2. As a result, the nest mortality of all species is lower also.

These observations are not unique. Open-nesting birds in the same place have, as a rule, the same nest-success rate, regardless of their own species' density (Nice, 1957; Nolan, 1963; Ricklefs, 1970). Nolan has noted that a seasonal rise in nest survival was shown by all the late-nesting species, whether relatively rare or abundant. Yet within- and between-habitat comparisons examined here indicate distinct density dependent effects.

PART II—THE BREEDING BIRD COMMUNITY

If all the bird species have rates of nest mortality similar to the field sparrow, then the pine-broomsedge habitat that was occupied at high densities by breeding field sparrows was overcrowded not only to the field sparrows, but to the other species breeding there as well. This suggests that a balance of winter mortality and breeding success might be determining the habitat selection and breeding distribution of other species as well. In fact, like the field sparrow, many species usually chose to nest where their chances of successful breeding were lower. Then it seems possible that their winter mortalities are lower in these habitats to compensate for their low breeding success. Therefore the dominance, winter food hypothesis which explains these relationships in the field sparrow might apply also to these other species.

With this suggestion of generality, a theory of bird species' densities will be developed. It will serve primarily to explain Table 6, a list of breeding species abundances in the pine-broomsedge habitat. It will then be used to make more general predictions and to reinterpret the observations of other researchers in this area.

TABLE 6. Census in pine-broomsedge

Field sparrow (*Spizella pusilla*)	150 ♂♂/100 a
Yellowthroat (*Geothlypis trichas*)	46 ♂♂/100 a
Cardinal (*Richmondena cardinalis*)	20 ♂♂/100 a
Towhee (*Pipilo erythrophthalmus*)	20 ♂♂/100 a
Prairie warbler (*Dendroica discolor*)	6 ♂♂/100 a
Indigo bunting (*Passerina cyanea*)	6 ♂♂/100 a
Blue grosbeak (*Guiraca caerula*)	6 ♂♂/100 a
Mockingbird (*Mimus polyglottis*)	3 ♂♂/100 a

Development of a General Theory of Breeding Bird Densities

Let n_{ij} = density of ith species * in jth habitat (breeding density only)

b_{ij} = per individual production of independent offspring minus percent breeding season adult mortality, ith species in jth habitat

g_{ij} = replacement rate of ith species in jth habitat

d_{ij} = proportion of individuals of ith species in jth habitat dying during nonbreeding season

$S_{ij} = g_{ij} - d_{ij}$

A_j = area of jth habitat

$N_{i\cdot}$ = population size of ith species

$N_{\cdot j}$ = density (all similar nesting species combined) in the jth habitat

b_i = per individual production of ith species

Now we consider some assumptions that are suggested by the earlier observations.

Breeding success. In the previous part it was suggested that the nest success of a species in a habitat is inversely de-

* Only small open-nesting bird species considered.

pendent on the joint or sum density of similar nesting species in that habitat. Thus, we assume $g_{ij} = f(N_{\cdot j})$ such that $N_{\cdot j} \uparrow \Rightarrow g_{ij} \downarrow$. Probably replacement rate depends on the habitat as well so that we can write $g_{ij} = f_j(N_{\cdot j})$. (This symbolic statement reads: the birth rate of the *i*th species in the *j*th habitat depends on the summed species densities in the *j*th habitat, in a way that depends on the *j*th habitat.)

Mortality. The hypothesis of winter food limitation and dominance-regulated exploitation implies that winter populations limited by food supply have density-dependent mortality, at least above certain densities. Once the (winter) density reaches the level that the habitat will support, any additional birds are doomed (or represent birds that are doomed) to starve. Thus, as the winter density increases above this maximum support level, individuals with the highest possible mortality rates are being added to the population, increasing the average mortality rate. Some question remains however, about a relationship between breeding density and survival.

Before formally stating any assumptions, we must consider the relationship between breeding density and winter density. If the net birth rate does not decrease too fast with an increase in breeding density (i.e., the transformation curve is not folded), then an increase in breeding density will result in a larger population at the beginning of winter and higher winter densities. However, if there is a large enough decrease in birth rate with an increase in breeding density, the net production of offspring may be so reduced that the added number of adults cannot compensate for it. Then there will be a reduction in the total number of prewinter individuals. In this case, an increase in breeding density results in a reduction in winter density and thus, in winter death rate. We shall, for simplicity, exclude this case in future arguments. This exclusion is partly justified since we are elsewhere assuming limitation by winter food supply, and a strongly negative birth rate-density relationship would be realized only in populations limited mostly by breeding habitat. However, as I have pointed out earlier, a population can be limited by both wintering and breeding habitats. The exclusion must therefore be regarded as restricting application of the resulting theory to species limited largely by winter resources. It permits us, however, to assume that winter mortality is always positively related to breeding density, expressed symbolically in the following equation: $d_{ij} = f_j(n_{ij})$ such that $n_{ij} \uparrow \Rightarrow d_{ij} \uparrow$.

This is not all that we can assume, however. It is possible that in the winter there is mixing of birds from different breeding habitats, so the winter densities, and therefore winter mortalities of birds from one breeding habitat may be affected by the breeding density in another. This is in fact what happens to field sparrows which breed in nonwintering habitat types. These individuals naturally winter in wintering habitat types where they compete at a disadvantage with field sparrows which have bred in those habitats. Presumably, when more of the latter individuals are present, more immigrants suffer. Hence, the mortality of birds breeding in nonwintering habitats is affected by breeding densities in wintering habitats. We assume, in general that $d_{ij} = f_j(n_{ij'}; j' \in L')$, $n_{ij'} \uparrow \Rightarrow d_{ij} \uparrow$ where L′ is the set of all habitats occupied by the ith species ($\in$ means "is a part of"). Thus, we assume that the death rate of all the birds of a species breeding in a habitat increases with the species density in any of the habitats occupied by the species. All habitats must be taken into account simultaneously.

Habitat densities. Let us assume that each species has some fixed population size. Suitability (S_{ij}) as defined above depends inversely on death rate and positively on birth rate. Thus, for each species in each habitat it is a decreasing function of each of the separate species densities in that habitat, as well as of the densities of that species in all of the habitats occupied by it. We can write

$$S_{ij} = f_j(n_{ij'} : j' \in L_i'; n_{1j}, n_{2j}, \ldots, n_{Ij})$$

where $n_{ij'}, n_{1j}, n_{2j}, \ldots, n_{Ij} \uparrow \Rightarrow S_{ij} \downarrow$. For each species, an ideal distribution is achieved when suitabilities are equal in all occupied habitats and greater than the maximum of any unoccupied habitat (see Chapter 4). If there are m' occupied habitats, then equality in all of these yields $m' - 1$ equations of the form $S_{ij'} = S_{ij''}$ where $j', j'' \in L_i'; j' \neq j''$. Since the total number of habitats is m, then $m - m'$ must be unoccupied. There are $m - m'$ equations of the form $S_{ij} = S_{ij(\max)}$, $j \in m' - 1$. This yields a total of $m - m' + (m' - 1)$ or $m - 1$ equations for each species. There are I species, thus there are $(m - 1) \times I$ of these suitability equations plus I population size equations ($N_i = \sum_{j=1}^{m} A_j n_{ij}$). These yield $m \times I$ total equations in as many density variables. Hence, since all the functions involved are monotonic *re* their arguments, unique, ideal distributions (sets of densities) are determined at every population size for every species.

The equilibrium population size can be determined for each species by the conditions of equal birth and death rates. These rates are $g_i = b_i/(1 + b_i)$; $d_{i\cdot} = \left(\sum_{j=1}^{m} A_j n_{ij} d_{ij}\right)/N_i$. (both monotonic in the densities). The equilibrium, as shown in Chapter 1 is $g_i = d_i$. There are n such equations which, with the equal suitability equations above, yield enough equations to determine a unique set of densities, and of course, population sizes.

Note that these are "ideal" solutions as defined and restricted in Chapter 4. As such, they represent an evolutionary ideal, useful only as an approximation of reality. The determination of the equations is a well defined (if enormously difficult) task. This theory thus suggests an understanding or definition of the problem of species densities without offering any sort of actual solution.

Application of the Theory to the Pine-broomsedge Habitat

We can better visualize what this definition means if we do some graphical simplification. We first reduce the problem to two dimensions. This can be done by considering variations in the abundance of each species when all others are fixed at equilibrium. At each population size of the one variable species there is a unique distribution. Associated with this distribution is a unique set of birth and death rates and suitabilities which can be graphed versus the separate densities in each habitat on two-dimensional graphs.

Insofar as species population sizes are independent, the resulting graphs are a sort of average of what would be measured in practice. If population sizes of different species are not independent, somewhat different curves would be obtained which are quite difficult to define mathematically. However the curves are defined, the same equilibrium points are determined.

Now consider Figure 46. The upper curves are replacement-rate curves for three different species in a particular habitat. The lower curves are the death-rate curves for the different species in the same habitat, as defined above. Although the nest mortality in the three species is expected to be similar, their replacement rates will be slightly different due to differences in clutch size and length of nesting season. As indicated above, the replacement-rate curves are dependent on the sum density in the habitat. Thus, the

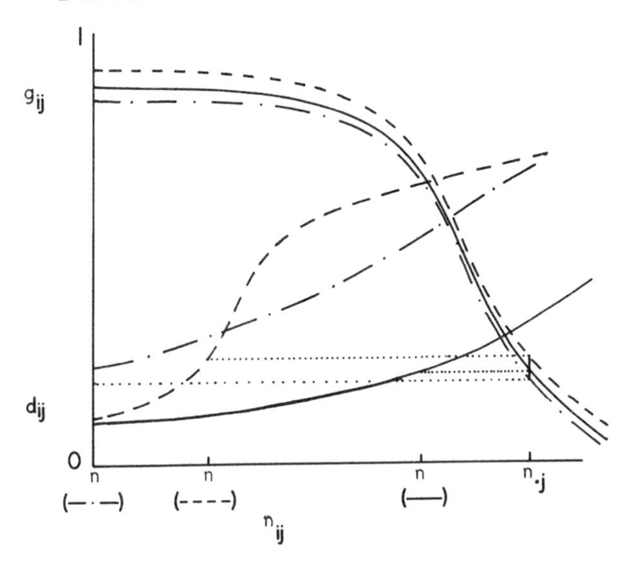

FIGURE 46. Regulation of densities in a three-species community. Birth-rate and death-rate curves for three species are given (solid, dashed, dot-dashed). The dotted lines are equilibrial $g \leqslant d$ values that satisfy the theoretical restrictions: (1) they intersect the g curves (declining) on some vertical line (drawn in solid at $(n_{\cdot j})$, and (2) they intersect the d curves (rising) at separate species densities, n_{ij}, such that $\sum_i n_{ij} = n_{\cdot j}$. The lowest dotted line connects $g \leqslant d$ for the dot-dashed line species (here, $g < d$, $n = 0$). The middle dotted line connects $g \leqslant d$ for the solid line species (here $g = d$ at a rather high n). The top dotted line connects $g \leqslant d$ for the dashed line species (here $g = d$, at a lower $n > 0$).

permitted combinations of species replacement rates are determined by the sets of intersections of all vertical lines with the three curves. The sum density in a habitat has a fixed value and affects all the species present simultaneously. The equilibrium values of density at equal replacement and death rates in the habitat are demonstrated by the three horizontal dotted lines connecting the death-rate and the replacement-rate curves. These densities are determined by the intersection of these horizontal lines with the death-rate curves. They are given on the n_{ij} axis. The equilibrium densities must sum to $N_{.j}$, the sum density. This restriction can be satisfied by only one set of horizontal lines and intersections.

It is easy to see that some species have higher equilibrium densities than others and that the difference is dependent for the most part on the death-rate curves. A species with a death-rate curve initially low and broad (solid lines in Figure 46) is apt to have a higher equilibrium density. A species with an initially low but rapidly increasing death-rate curve (dashed lines in Figure 46) will have a low density. A species with an initially high death-rate curve (dot-dashed lines in Figure 46) will have zero density. We ask, what do these differences between curves mean? A species, such as the field sparrow in the pine-broomsedge habitat, which gains a considerable experience or dominance advantage to enhance winter survival should have an initially low mortality curve in accordance with this advantage. Top dominant field sparrows probably survive at an annual rate of .80 versus .50 for the population average. Furthermore, the field sparrow winters in such high densities in the pine-broomsedge habitat (up to ten birds per acre) that a considerable number of individuals can obtain this dominance advantage and still benefit. For instance, if the habitat supports five birds per acre, then only the five top dominants per acre in that habitat can survive. If six birds per acre attempted to breed in the habitat and thereby become top dominants, one bird would not be able to receive the benefits of his experience and dominance. He would be subdominant to the other five and as doomed to starve there as any other individual subdominant to the top five. Since relatively many field sparrows can benefit from breeding experience, the mortality curve not only should be initially low, but also should stay low up to a high density. As noted, this type of curve should be associated with a species that maintains a high breeding density. This is what we observe in Table 6.

Let us consider another species, the towhee (*Pipilo erythrophthalmus*). This species also winters in the pine-broomsedge habitat but at considerably lower densities than the field sparrow (about .5/acre, unpublished observation). Like the field sparrow, it too might well receive some benefit from breeding in pine-broomsedge, the winter-type habitat, but only relatively few individuals can so benefit because of the low winter densities. Thus, the towhee mortality curve should start low but should rise even at low densities. This expectation would result in a breeding density which is low but greater than zero. Again, this is what is observed.

The junco (*Junco hyemalis*) also winters in the pine-broomsedge habitat but does not stay to breed. This species, however, in its winter ecology, occupies a less stable part of the habitat than do either the field sparrow or the towhee. The broomsedge tussocks used by the field sparrows are, in an annual cycle, an almost unchanging aspect of the pine-broomsedge habitat. The dense pine clumps beneath which the towhees feed are also relatively stable. We may reasonably presume that habitat stability is an important part of the advantage gained. Without it, breeding experience would not correspond with winter needs. The junco, in the winter, usually feeds in the patches of tall weeds in these habitats, but these patches disappear in the spring and are often replaced by broomsedge (see Oosting, 1942, on plant succession). New patches grow up again later in the summer, but these patches, during the normally early breeding season, do not have a similar appearance to the winter tall weed patches. Thus, juncos staying to breed in the pine-broomsedge habitat could get little appropriate experience to aid in winter survival. Therefore, lacking any advantage, they should have a mortality curve which is close to the species average and is, therefore, relatively high, even at low densities. Presumably, the initial value is so high that it exceeds the potential birth rate for juncos in that habitat, so the species does not breed there. (Of course, in northern habitats, no species can winter and the habitats are not filled by winter residents. These are the habitats available to species like the junco.)

The birth rate for the junco is, of course, low due to the high densities of other species (Figure 46), especially the field sparrow, which, with its dominance advantage, is able to tolerate the resulting low replacement rates. The field sparrow (with other species) thus raises the joint species density to such a level that the potential junco replacement rate is reduced below the junco minimal death rate. Field sparrows (and other species) thereby exclude juncos from breeding in this habitat.

It is important to remember that the average survival rate of a species is an average of many different expected values. The top dominant junco, like the field sparrow, probably survives a winter season with a probability of nearly .80 (Fretwell, 1969a), but the average population survival rate is much lower (about .50, as for most other species). However, the top dominant field sparrow will probably come from a breeding habitat similar to the wintering habitat (e.g., the pine-broomsedge habitat). This is perhaps true of the top dominant junco as well, but the breeding habitat similar to the wintering habitat for the junco is not the pine-broomsedge old field. Our hypothetical juncos breeding in old fields will be either subdominant or of average dominance status depending on whether other individuals of the species can find breeding habitats more similar to their wintering grounds. If such breeding habitats are occupied, juncos from them will be more dominant than a possible old field resident.

The yellowthroat (*Geothylpis trichas*) breeds in the pine-broomsedge habitat but winters in coastal marshes to the south and east. According to the above explanation, this species must in some way be able to get experience which is advantageous in winter survival by breeding in this habitat. Perhaps the dry Johnson grass (*Sorghum halepense*) stands and dry broomsedge (*Andropogon virginicus*) tussocks which are present through the breeding season are sufficiently similar to a winter marsh to provide the appropriate breeding experience. It certainly must be possible for migrants to carry advantageous breeding experience to their wintering grounds, since the demonstration of such advantage was in part conducted on populations of migrant field sparrows (Fretwell, 1969a).

These explanations are summarized in Figure 47, where the major breeding and wintering species of pine-broomsedge habitats near Raleigh, North Carolina are given hypothetical mortality curves, and for simplicity, a common, equally hypothetical birth-rate curve. The mortality curves are based on winter densities, similarity between the pine-broomsedge habitat, and observed breeding densities. The dotted line establishes the unique equilibrium densities. Its intersection with the replacement rate curve yields the joint density, $N_{.j}$. The intersections with the death-rate curves determine the separate species densities, which sum to $N_{.j}$. Species whose mortality curves start above this horizontal line (like the junco) are effectively "out-competed" for the breeding space and do not breed in the habitat. Curves similar to that for the junco could be drawn for all other species that do not breed in the habitat.

This completes the application of the theory to the breeding bird community of the pine-broomsedge habitat. The essential points of this application of the theory are that birds select a habitat for breeding on the basis of similarity to winter habitat type and average nest mortality.

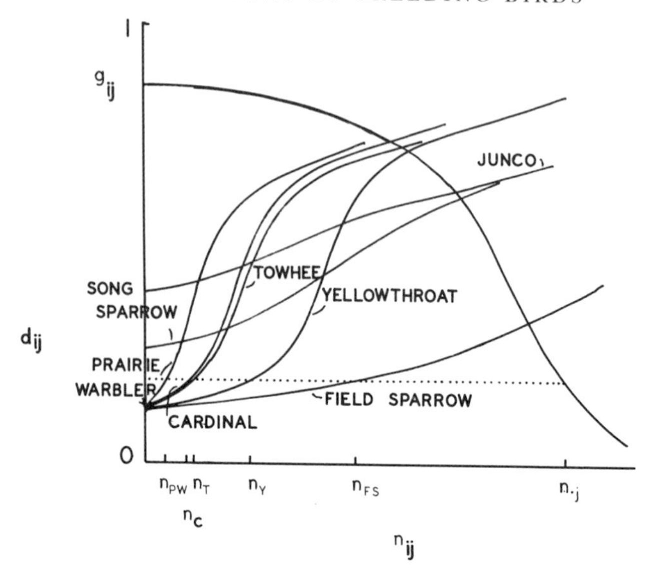

FIGURE 47. The pine-broomsedge avian community, some hypothetical curves of density regulation as in Figure 46, except that only one g curve is drawn, and all species have the same dotted horizontal line connecting $g \leq d$. The text explains why species have different d (rising) curves.

The relative breeding densities are determined by relative winter densities, the sum density by nest predation. In the following paragraphs, this theory will be used to reinterpret previous theory and to reanalyze previous data.

The objective in this extension is not to assert that all breeding bird communities are similarly regulated, but to explore the extent to which these ideas help us understand other communities. There will surely be habitats (e.g., marshes?) where the ideas have no significance.

Application of the Theory to Other Theories and Data

The "broken stick." The assumption in Figure 47 that all open-nest, small-bird species have a common replacement rate is probably not too far from reality. At least all of the species should show a sharp drop in replacement rate at approximately the same joint density, since the same predators are probably involved in the nest mortality. (Of course, hole-nesting species would probably require different sets of curves.) Thus, the joint density in the habitat is an approximately fixed value which is determined by the density at which certain of the predators decide that it is profitable to prey on nests of birds. The species in the habitat are limited by this joint density. This provides an interpretation of MacArthur's (1957) broken-stick model of bird communities. The "stick" is the dotted line in Figure 47, and it is, in effect, "broken" into nonoverlapping pieces.

The role of winter food in the distribution and abundance of breeding birds. The present theory has assumed that the bird populations under consideration were limited in abundance by winter food. The theory has attempted to describe the factors (especially nest mortality and breeding experience) which regulate the breeding distribution of such populations. There are two alternate hypotheses. It is possible that the total species populations are limited in abundance by breeding resources, not winter food. It is also possible that breeding food supplies, not nest mortality, limit the breeding distributions. These alternate views find no support in my studies of birds in the pine-broomsedge communities.

In the pine-broomsedge habitat, the field sparrow comprises about 55 percent of the total population; the yellowthroat and prairie warbler together make up almost 25 percent. All these species feed on insects in the breeding season. Under the hypothesis that energy is limiting, it is necessary to postulate that there is some source of insect energy that the field sparrow is more efficient at harvesting than either the yellowthroat or the prairie warbler. This is hard to accept. The field sparrow, because of its heavier bill, seems better adapted to a seed diet while the two warblers seem better adapted to insect feeding. The warblers divide the two available vegetative layers between them. Food taken by field sparrows is so diverse (beetles, spiders, moths, moth larvae, grasshopper larvae, different pairs specializing on different foods, unpublished data), that comparisons with the warbler species on this basis is not justified. The field sparrow in the breeding season evidently takes what is there to take. One wonders why, if breeding food is limiting, the warbler in that niche does not take it first.

It is still possible, however, that the breeding distributions of these species are determined by food. Individuals might breed where food is more abundant, going elsewhere only after density-dependent competition for food becomes critical. However, it does not seem to be important in the successional stages studied. The nest mortality in these habitats was very density-dependent, causing 100 percent changes in production (.44 broods per pair versus .89 broods per pair). Changes in brood size, which might reflect competition for food, were much smaller (3.2 to 3.6, about 10 percent). Perhaps food is more important in forests where there are more hiding places for nests, or fewer nest predators, or food may be more important to hole-nesting birds.

Migration. A corollary of the theory is its statement regarding the adaptive value of migration from wintering grounds. Assuming limitation by winter food resources, it is quite clear why some wintering field sparrows migrate from the pine-broomsedge habitat to breed. The highest winter densities in this habitat type are about twice the highest breeding densities. If all the wintering birds stayed to breed, the nest mortality rate would be too high for successful replacement, in fact very close to 100 percent (see Figure 45). Some birds thus find it advantageous to leave, to go 1,000 yards or 1,000 miles, to areas unoccupied in winter by field sparrows or by any other similar nesting species in large numbers. They forsake the dominance (experience) advantage. Their life span is shorter but they raise more young, which balances this disadvantage.

A possible explanation for the total migration of other species from the wintering grounds is offered, as in the case of the junco. Also, the migration of the yellowthroat into the pine-broomsedge habitat is tentatively explained. These phenomena are attributed to similarity between winter and breeding habitat types. This is in agreement with MacArthur (1959), who in a general survey of migratory populations suggested that the stability of the vegetation was involved. MacArthur was more concerned with stability of the breeding habitat, but of course the changes from summer to winter are the inverses of the changes from winter to summer. Unstable habitats should change their avifauna with the seasons, and they do.

Competition in the breeding season. Gause's axiom asserts that identical species cannot coexist in the same habitat and that different species can occupy the same functional niche only in different places. Figure 47 can be regarded as a formal disproof of this axiom insofar as it is applied to just one season. For in the theory of Figure 47, all the breeding birds of the pine-broomsedge are *assumed* to be *essentially* identical in their breeding ecology, yet coexistence is easily established by winter differences. Thus, there is no *a priori* reason to expect differences between coexisting breeding bird species, which may be as similar as they like.

Predictions from the Theory

In this section, some predictions from the theory will be stated and tested.

Foliage height diversity (FHD) and sum abundance. The present theory argues that the sum density of open-nesting birds is limited by nest predation. We should therefore expect that open-nesting birds increase in density with some measure of nesting cover, for example, the number of layers of vegetation or FHD. We do not expect this to be true of hole-nesting birds except where FHD is correlated with the availability of holes.

These predictions are tested in Figure 48, using data from MacArthur and MacArthur (1961). Cover is indexed

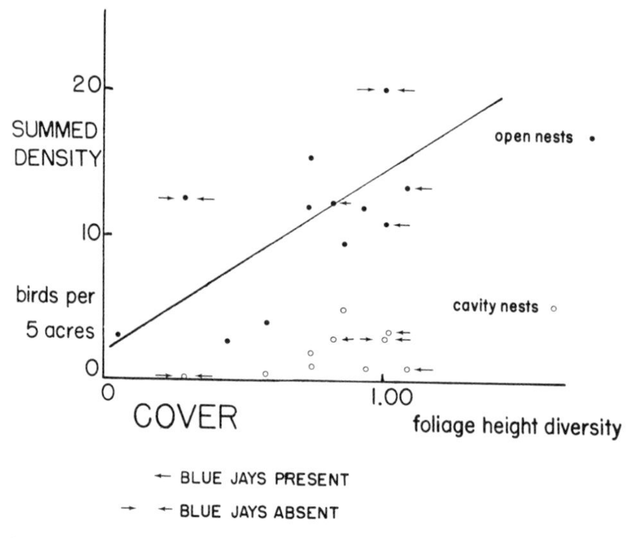

FIGURE 48. The effect of cover on the summed density of open-nesting and cavity-nesting (hole-nesting) species of birds. Foliage height diversity, or the number of layers of vegetation, is taken as the measure of cover. There is a significantly greater increase of open-nesting individuals with an increase in cover, compared with the increase of cavity-nesting pairs. Also, the open-nesting increase is significant statistically; the cavity-nesting trend is not. Arrows mark the presence or absence of blue jays, a nest predator. Open-nesting densities are higher where blue jays are absent, but cavity-nesting densities are unaffected (see Figure 6). (Data from MacArthur and MacArthur, 1961.)

by FHD. The hole-nesting sample excludes sets of data from which hole-nesting birds were absent, as holes may not have been available. The increase in sum density of open-nesting species with FHD is significantly larger than the increase in sum density of hole-nesting birds. Also, the first correlation is significantly positive, the second is not.

Blue jays (Cyanocitta cristata) and sum density. The theory predicts that the presence of blue jays, a major predator of open nests, should significantly diminish a habitat's suitability to open-nesting birds. Then we should expect lower sum densities of open-nesting birds in woods where jays are found. This prediction is not expected to hold for hole-nesting birds.

In Figure 49, the sum density of open-nesting and hole-nesting birds is plotted versus the density of blue jays. The data are from Audubon Field Notes. Closed circles are from a single woods (Rock Creek Park in Washington, D.C.). The open circles are all from upland deciduous censuses from 1967. All the points are in essential agreement. The sum density of open-nesting species significantly decreases with an increase in jays while the hole-nesting sum density increases. The increase of the latter seems lower and more variable than the decrease of the former.

Figure 49 confirms the expectation of the theory. But the fact that open-nesting and hole-nesting birds are inversely correlated suggests an alternate explanation, namely that the two groups compete for food, and increases in one result in decreases in the other. This explanation does not account for the fact that the blue jay, which is an open-nesting species, increases with the hole-nesting species. Also, hole-nesting and open-nesting abundances were not correlated in the preceding analysis.

The present theory would interpret the positive correla-

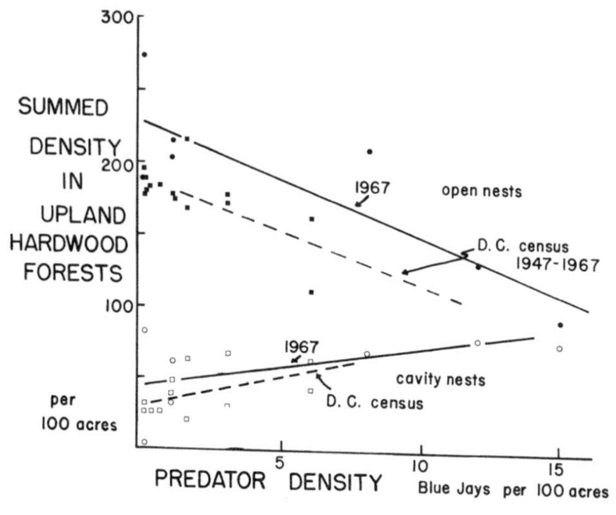

FIGURE 49. The effect of nest predators on the summed density of open-nesting and cavity-nesting species of birds. The density of blue jays is taken as a measure of predator density. The summed density of open-nesting species declines as the number of blue jays increases, both between upland forest habitats in the same year, and within a single upland forest habitat over different years. There is a smaller increase of cavity-nesting pairs with an increase of blue jays. (Data from Audubon Field Notes breeding bird censuses.)

tion between hole-nesting and blue jay densities as evidence that the two groups are ecologically similar in the winter, when their populations are presumed to be controlled. In fact, the hole-nesting species and jays are ecologically similar in that both tend to be permanent, year-round residents. Only a few hole-nesting forest species migrate. The majority (the tits and woodpeckers) are permanent residents like the blue jay. Events likely to favor the survival of one permanent resident (high winter temperatures, a good mast year, establishment of winter feeding stations) may well enhance survival of others. So abundance of jays may be correlated with abundance of other species. I especially wonder about the role of winter feeding stations in this regard. Near my home in central New Jersey, the extensive winter feeding by neighbors appears to have resulted in an extraordinary density of breeding blue jays. No less than seven pairs bred in less than two acres of woodland! Three nests in my own yard were less than 30 feet apart. Many of the areas censused in the Audubon Field Notes are perhaps similarly affected by nearby feeding stations.

Rosenzweig (pers. comm.) has suggested that limitation by predators of open-nesting species releases resources for hole-nesting species which therefore increase in density. He observed that three open-nesting pairs are lost before one hole-nesting pair is added (compare slopes on Figure 49). It may be significant then that the clutch sizes of hole-nesting species are about 1.5 times the clutch sizes of open-nesting species (Skutch, 1949), while survival of nests is twice as great (Nice, 1957). Thus hole nesters use $1.5 \times 2 = 3$ times the resource of open nesters, exactly the difference in slopes of the curves in Figure 49.

These findings are in agreement with Udvardy (1957), who noted that oak-hickory forests had fewer birds than beech-maple which had fewer than mixed hardwood (not dominated by oaks, hickory, or beech trees). Blue jay populations are probably limited by winter food, usually acorns and hickory and beech nuts. Jays should occur most frequently where their winter food is found. However, Udvardy noted that wet bottomlands had the most individuals of all, including some jays. I attribute this to the difficulty mammalian and reptilian nest predators have in finding and preying on nests located over water, but this requires further study.

Species diversity and winter densities. A pair of predictions about species diversity can be made. Since the theory requires that the total bird density is essentially fixed by predation, if a habitat contains species with high densities, there can be only a few species. The theory also says that relative breeding densities are largely determined by winter densities which are in turn determined by winter food. If food is limiting, there are two factors affecting winter densities: the abundance of food and the food requirements of the individuals. Species which are vegetarian in the winter, since they are on a lower trophic level, would on the average have a more abundant food resource and would therefore have higher winter and breeding densities. Species of small body size, since each individual needs less food, would also probably be able to maintain higher winter densities. Since high winter density implies high breeding density implies low breeding species diversity, we predict that the number of breeding (summer) species in a habitat is positively related to the proportion of individuals that *in the winter* are not vegetarians and to the number that are large.

In Figure 50, I have plotted the number of species found in censuses (from Audubon Field Notes) of 15-acre plots versus the winter density index, averaged over individuals found in the habitat. The species number was corrected for numbers of individuals in the census plots. The winter density index for each bird was the simple sum of two other indices; one of body weight and one of trophic status. The index of winter trophic status was calculated as follows: Fringillids, Mimids, grackles, and doves were regarded as strictly vegetarian and given the value 1. Warblers (except the myrtle warbler), flycatchers, vireos, kinglets, and creepers were regarded as strictly insectivorous and given the value 2. All the others were given the value 1.5, as their diet probably is mixed. These index values were averaged over all the open-nesting individuals in the habitats.

The index of body size was derived by giving the value 1 to all individuals less than 15 grams (field sparrow and

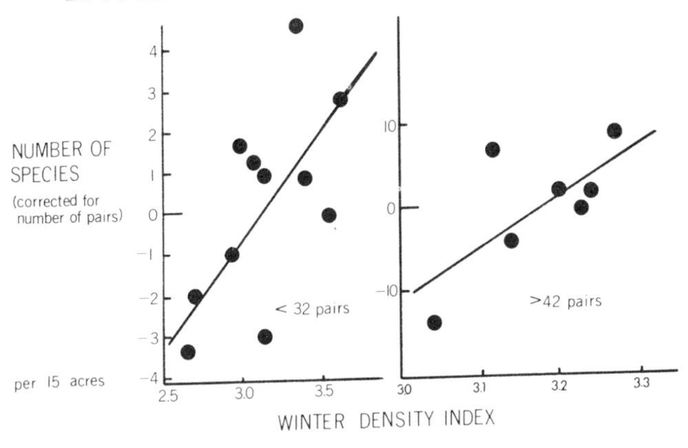

FIGURE 50. Winter density and breeding species diversity. The winter density index (WDI) is computed from the average body weight of the birds breeding in a habitat, and the average *winter* trophic status. Birds were classified into three logarithmic weight classes and three trophic classes (vegetarian, insectivorous, and mixed), and the average ranks were obtained for each habitat. The trophic average was added to the weight class average to obtain WDI. Winter density should decrease with an increase in body weight, due to increased food requirements, and should decrease with an increase in trophic status as insectivores have a sparser food supply than vegetarians. This assumes that winter populations are limited by food supply. Then higher values of the WDI represent populations that, on the average, have *lower* winter densities. The hypothesis predicts that breeding densities are regulated by winter densities, but that the summed species density is fixed by nest predation. Species with high winter and breeding densities (low WDI) take up more of the available space in the breeding habitats, and permit fewer species to settle. The data in the graph are taken from Audubon Field Notes, breeding bird censuses, 15-acre plots only. After correcting for number of pairs in the habitats, there is a significant increase in number of species with an increase in WDI. The trend seems to be different in habitats with many pairs, compared with habitats with few pairs.

smaller); 2 to all individuals between 16 and 35 grams (junco to white-throat) including the catbird, and the thrushes except the robin (*Turdus migratoris*); and 3 to all larger individuals. The habitat average of the body size index values was then obtained for open nesters.

A multiple-regression analysis found that both of these

variables explained a statistically significant amount of the variation in number of species and that the number of species increased by about seven for a unit increase in either index. This is qualitatively as predicted from the present theory. If breeding birds are limited solely by energy in the breeding habitats, *winter* trophic status could well be irrelevant.

Foliage height diversity (FDH) and bird species diversity (BSD). MacArthur and colleagues (1961, 1963, 1964) have noted that the number of breeding bird species in a habitat is correlated with the foliage height diversity. This correlation is explained by the present theory as follows. By the argument predicting the results in Figure 50, the number of species is lower in habitats inhabited by a large number of vegetarian species. Vegetarian birds usually eat seeds or fruit. Seeds or fruits are produced by plants for reproduction and dispersal. We may expect that plants in unstable, successional environments put more of their energy into producing these elements of reproduction and dispersal than plants in stable, climax habitats. This has been found to be true (Salisbury, 1942). The effect of this on bird populations is documented in the data of Quay (1947) where the number of seed-eating birds is found to be high in successional habitats and low in climax or near climax habitats (Figure 51). Also plotted in Figure 51 is the average species density in each of the habitat types. Note the much higher densities in early successional stages.

This being the case, and since the present theory requires birds to breed in habitats similar to the wintering habitat, we should expect more vegetarian, high density species to breed in successional habitats. Since in breeding habitats, more vegetarian individuals are presumed to result in a lower number of species, successional habitats should have fewer species. Successional habitats also have

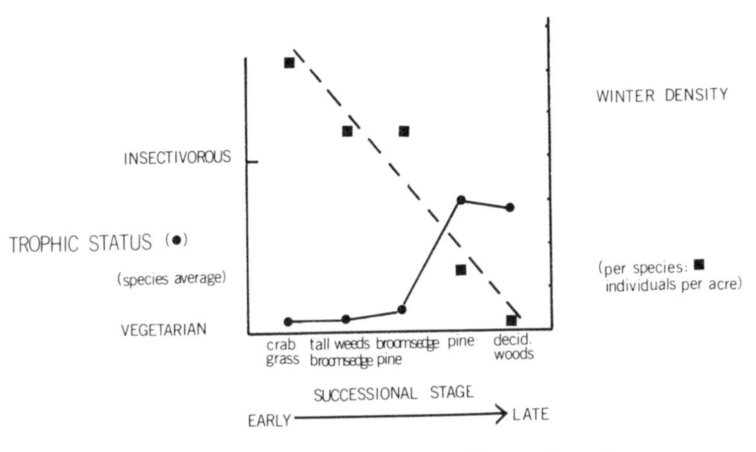

FIGURE 51a. Successional stage, winter densities, and breeding species diversity. Climax and subclimax successional stages are occupied in winter by insectivorous species (●) which have, on the average, very low densities (■). The hypothesis predicts that these species will also have low breeding densities, so more of them can breed together in habitats where the total density is fixed by nest predation (Figure 51b). The breeding species diversity will be higher in climax habitats since birds tend to breed in habitats similar to the wintering habitat. Climax habitats have more vegetative layers, hence a higher foliage height diversity. Therefore, there is a correlation between foliage height diversity and breeding species diversity (MacArthur and MacArthur, 1961). Note that a *climax* spruce forest with a *low* FHD had many insectivorous species, as this theory would predict. This high diversity breaks up the FHD-BSD correlation in MacArthur and MacArthur, 1961. However, the low FHD of this habitat might be misleading as layering in a spruce forest is different. This habitat also had more pairs than its FHD would predict (Figure 48, leftmost double-arrow point above the line).

a lower foliage height diversity, hence bird species diversity and foliage height diversity should be correlated, as MacArthur has already found. It is encouraging to note that one of the points not fitting this correlation in MacArthur and MacArthur's (1961) analysis was a low foliage height diversity *climax* spruce forest which had more species (most purely insectivorous) than would be expected on the basis of its FHD. Of course, MacArthur and MacArthur's ex-

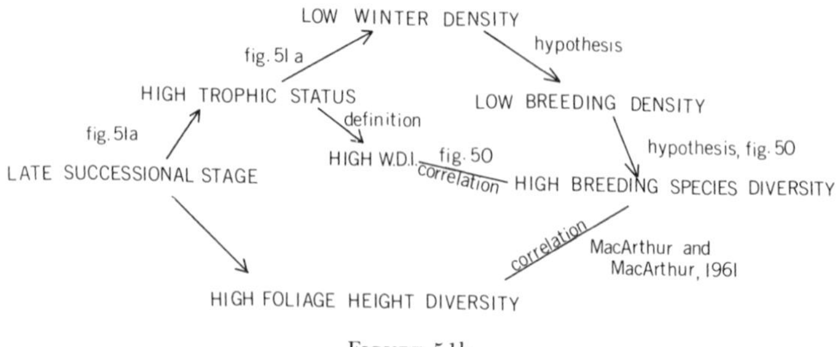

FIGURE 51b.

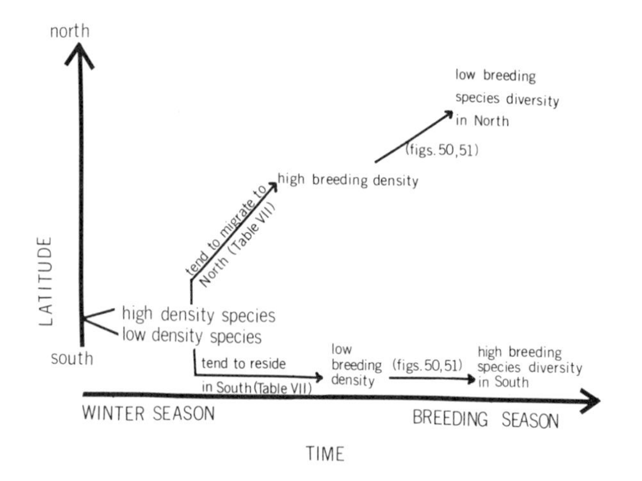

FIGURE 51c. Latitudinal gradients in bird species diversity in the breeding season. In a given wintering habitat, there is a tendency for the species wintering at high densities to migrate, while the species wintering at low densities are more often permanent residents (see Table 3). Therefore, the northern habitats, in the breeding season, are more frequently inhabited by (migrant) species of high winter density. Then, by the arguments in a and b, the breeding species diversity in northern habitats will be lower. Cox (ms.) has demonstrated that in fact breeding communities with more migrants *are* less diverse. Skutch (1954) has shown that birds breeding in the tropics (which are almost all permanent residents) have lower breeding densities than birds, usually migrants, at higher latitudes.

planation of this wayward point is quite good—spruce forests are *not* layered horizontally, and so FHD is not well measured.

Latitudinal gradients in bird species diversity. This theory provides yet another explanation of the latitudinal trends of bird species diversity (Pianka, 1966; MacArthur, 1965). In general, the theory suggests that if there are more breeding species in tropical habitats, this effect must be due to lower wintering densities of tropical-breeding birds.

Two things differentiate tropical-breeding birds from temperate-breeding birds: (1) the former are more often permanent residents, and (2) resident or migrant, tropical species usually *winter* farther to the south. This suggests two possible hypotheses. (1) Communities of migrants might have lower diversities than communities of residents; then tropical communities, with more residents would have a higher diversity. (2) There might exist a latitudinal decrease in *wintering* bird species diversity for residents or migrants or both; then tropical communities which have wintered farther south would be populated from richer winter communities. In terms of the present theory, we should expect from these possibilities that (1) since high winter density and migration both imply low breeding diversity, migrants must winter at higher densities than residents; (2) relatively tropical-wintering populations must have a lower density per species than temperate-wintering populations. This last expectation will certainly not be absolutely valid in that wintering birds of very high latitudes, where there are very few residents at all, will have relatively low densities.

The prediction (1) that migrants winter at higher densities than residents might be explained as follows. We have argued that where a species breeds depends on where it can obtain experience that will benefit winter survival. If a species is highly territorial in the winter, then the relevant experience will be highly specific, even to familiarity with certain rocks, trees, and so on. Such a species would find advantage in breeding in the exact site of its winter territory and would thus be inclined *not* to migrate. If a species wanders about a great deal the relevant experience is of a habitat type; certain structural features, plant forms, and so on. Should an empty breeding habitat exist which is similar to the wintering type of some wandering species, the wandering species might well migrate there. The experience advantages so obtained would be as appropriate as that obtained from breeding on the specific wintering grounds.

Birds which flock are more generally wanderers, so tendency to flock should be correlated with tendency to migrate. If tendency to flock were also correlated with density, then density *per se* would be correlated with migratory activity. It remains then to ask why density might be correlated with flocking.

There are two reasons why we might expect birds wintering at high densities to flock. Flocking is in part an adaptive response to predation which provides cover and increased guardedness for the members (Allee et al., 1949, but see Cody, 1971). Such cover would be especially advantageous to species feeding in open areas without sufficient vegetative cover and to species feeding at such high overall densities that predators might be attracted. We can therefore conclude generally that high winter density and/or feeding in an open environment should usually be related to flocking.

Now note Quay's data (Figure 51a) which show winter density to be higher in open habitats. This means that if openness of habitat leads to flocking, then flocking is again related to high density. Thus, we can generally hypothesize a relationship between density and tendency to travel in flocks. And since flocking can be expected to be correlated with migratory habit, we expect a similar correlation between density and migration.

There are many steps in this argument. The correlation between flocking and migration has not been demonstrated. However, it does provide another basis for the connection between winter density and migratory habit that is hypothesized to exist purely on the basis of latitudinal gradients in species diversity.

It is pleasing to find some of these arguments and predictions verified in observation. Cox (1968) has shown that a higher proportion of migrants in a breeding community is correlated with a reduction in its diversity. Quay (1947) has noted that half of the migratory species wintering in agricultural or natural habitats in Piedmont, North Carolina (8 of 16) were in the three highest abundance classifications, but that less than a third (8 of 26) of the residents were so common. A breakdown by habitat of Quay's data is even more impressive (Table 7). In pine woods, for example, only one of the five most abundant wintering species was resident, but all of the four least abundant species were resident. In pine-broomsedge only one of the four most abundant wintering species was resident while the two least abundant were both resident.

In the crabgrass habitat, however, one of the two most common wintering species migrated, but three of the four least common species migrated opposite to the density-migratory prediction. In the bare field, both of the two most common species were resident while three common species migrated, as well as one of the three least common species. The bare field birds again do not support the theory. Overall, however, Table 7 supports the idea that more abundant species migrate.

The preceding arguments and observations generally support the following explanation for tropical increases in

TABLE 7. Winter density and migratory habit

Habitat	Relative abundance* low	medium	high	Total	Stability	Openness	Total
Pine	*m*§ 0(00)	2(50)	4(80)	6(46)	6	6	12
	p‖ 4	2	1	7			
Deciduous	*m* 1(33)	4(57)	2(100)	7(58)	3	4	7
	p 2	3	0	5			
Pine-broomsedge	*m* 0(00)	0	3(75)	3(50)	5	5	10
	p 2	0	1	3			
Tall weeds-broomsedge	*m* 1(50)	0	3(75)	4(67)	4	3	7
	p 1	0	1	2			
Crabgrass	*m* 3(75)	0	1(50)	4(67)	2	2	4
	p 1	0	1	2			
Bare field†	*m* 1(33)	3(100)	0(00)	4(50)	1	1	2
	p 2	0	2	4			
Total	*m* 6(33)	9(63)	13(70)				
	p 12	5	6				
Total‡	*m* 5(33)	6(52)	13(76)				
	p 10	5	4				

Data from Quay, 1947.
* Excludes cavity nesters, birds of prey, and the common crow.
† Agricultural habitat.
‡ Excludes bare field data.
§ *m* = number of migrant species. Number in parentheses is percent migrant species in the particular relative abundance class.
‖ *p* = number of permanent resident species.

bird species diversity. In tropical habitats, high winter temperatures and productivity result in high winter populations. All these wintering birds cannot breed where they winter, due probably to density dependent nest mortality, so some migrate. The ones wintering at high densities migrate more frequently, settling in the north at higher breeding densities. Therefore, habitats with more migrants have fewer species. These habitats are farther from the equator so there exists a latitudinal gradient in breeding bird species diversity. Of course, there are probably other components to the latitudinal diversity trends observed in nature.

Coloniality. It is interesting to note one other prediction, this time regarding breeding coloniality, which may be expected in bird species having large winter flocks. Such species depend in the winter on a social environment as opposed to a vegetative environment, and might thus tend to form colonies in the breeding season simply because this creates conditions similar to those on the wintering grounds. I offer such a prediction primarily for species like the dickcissel and redwing blackbird, which travel in enormous winter flocks. Strict coloniality has not been observed in either species, except as determined by a very patchy sort of habitat, but tendencies to colonization may yet be found. For example, redwings in Ohio clover fields may be clumped in parts of a homogeneous hayfield habitat. Dickcissels may occupy some hayfields and not others.

Predictions To Be Tested

The preceding predictions and their verifications tend to support the hypotheses on which they are based. There remains considerable room for doubt, however, which can be resolved only by more thorough investigations. Needed are studies of winter bird ecology, especially along the lines followed by Quay (1947, 1957). Also, nesting mortality in breeding bird communities needs to be more widely investigated. Censuses in the same area, winter and summer, are needed as well as banding studies to determine the residency of individuals.

The existence of competition for food in breeding birds needs to be tested. As noted above, old field species do not seem to be seriously competing for food in the breeding season. The field sparrow, which is poorly adapted for insect catching, is more abundant than two warblers and nest mortality is very high. If the same sort of analysis has any relevance to open-nesting breeding birds in general (as Figures 48 through 50 suggest), then we would have to conclude that breeding birds do not necessarily compete with one another at least in the usual sense of the word. They do compete in the mathematical sense as the density of each species decreases its own fitness (nest survival rate) and the fitness of other species. But there is no "together seeking." If this is true, how do we explain the differences in feeding behavior observed classically by MacArthur (1958), and since then by others, for example, Crowell (1962). First, we should note that this theory does not suggest that the competition argument underlying the studies by MacArthur and others is not valid; only that it may not apply to breeding bird populations. Presumably, it applies to wintering bird populations, and wintering individuals are competing with one another in the usual sense. They would be expected then to show the sorts of differences observed in breeding populations and would have morphologies and reflexes appropriate to these different activities. Then, in the breeding season, even though competition for food hardly exists, the birds may be expected to retain their winter feeding differences to a greater or lesser degree depending on where food is most easily found, where nests are located, and so on.

It is also possible that in habitats where nest predation is not important, the habitat distribution is regulated by food supply. For example, the Maine spruce forest studied by MacArthur (1958) does not have many jays and may be relatively free of nest predation, so the differences in feeding behavior among the warblers breeding there may well reflect summertime competition.

This explanation can be tested against the alternate competition for food hypothesis by examining the effect of one species on the feeding methods and efficiency of another. For example, perhaps one could eliminate the nest predators in an area and thus effectively raise the populations of the breeding birds' nests. This should result in smaller surviving broods or lighter fledglings, if there is competition for food. One could also eliminate just one of a pair of competitors. Then if there is competition for food, the remaining competitor should raise heavier young, or if there is normally any starvation, more young per brood. Also, in this second experiment, the remaining competitor should expand its feeding behavior.

An alternate kind of experiment would be to provide winter food for some open-nesting, year-round resident. Hopefully, this species, in the breeding season, does *not* compete for food with at least one open-nesting migrant breeding in the same habitat. Winter feeding ought to raise the breeding density of the resident, and should therefore raise arbitrarily the open-nest mortalities of both the resident and the migrant. The migrant, in the course of a few generations, ought to show a reduced population in this habitat. At least the migrants breeding in the experimental habitat ought to have a significant selective disadvantage relative to migrants breeding in other habitats. The chipping sparrow (*Spizella passerine*), a migrant, and the pine warbler (*Dendroica vigorsii*), a resident, in an eastern pine woodland are a pair that could be used. Pine warblers come readily to fatty-type baits in the winter, and their breeding density should be raised easily by winter feeding. There is probably no breeding season overlap in feeding with chipping sparrows. Another pair (or group) that could be used are the siskins which nest in the northern coniferous forest. They feed their young on seeds, are resident, come to feeding stations, and have open nests. The effect of raising the density of siskins in the breeding season on the distribution or success of migrant warblers would be instructive.

Of course, the hypothesis presented here suggests that there is little or no competition for breeding food and that any manipulations of density or predators will have little effect on the per bird capacity to produce. It will affect only the nest mortality.

SUMMARY

The theoretical ideas of the preceding chapters are applicable, as demonstrated by an analysis of distributions of breeding birds. The field sparrow seems to be limited in abundance by winter food. The breeding experience is also a major component in annual survival, presumably through its effect on social interactions in early winter. Thus, the suitability of breeding habitats is influenced by similarity or proximity to winter habitats. Density-dependent effects stemming from higher rates of nest predation at higher nest densities are also important, so a field sparrow selecting a breeding habitat must take density into account as well. The available evidence suggests that the actual distribution may be close to the ideal free form unrestricted by territoriality. Field sparrows have higher rates of nest mortality in densely occupied habitats, but this lack of breeding success is compensated for by higher adult survival.

There is evidence that the field sparrow system is more or less general in that (1) the distribution of many other species may follow the same general pattern and (2) the nest mortalities of other species appears to be similar to, if not identical with that of the sympatric field sparrows. Thus many other species besides the field sparrow are most abundant in fields where their nesting success is lower, indicating habitat differences in adult survival. When these ideas are incorporated into a general theory, many verifiable predictions are generated. Migration is predicted to be correlated with habitat stability, sum abundance of open-nesting (but not hole-nesting) species should increase with cover, open-nesting (but not hole-nesting) species should decrease with nest predators (especially blue jays), species diversity should decrease with winter trophic status (i.e., breeding habitats occupied by species which are vegetarian in winter should have fewer open-nesting species). All of these predictions are verified. The theory also predicts that, given latitudinal gradients in bird species diversity, species which are more abundant in the winter should migrate more frequently. This, too, conforms to the data. The theory is consistent with other work on breeding bird communities.

There are more predictions as yet completely untested, but the evidence in hand suggests that the winter limitation—habitat experience—nest predation hypothesis is at least as plausible as any alternate hypothesis.

CHAPTER SEVEN

Winter Densities of Fringillids

INTRODUCTION

Building *suitability-density* models for a theory of habitat-distribution can be approached two ways. We can construct a hypothetical model of *a priori* form with unspecified constants, as was done at the end of Chapter 3. These models can be fitted to data to estimate the constants. As an alternative, however, we can attempt to construct the models entirely *a priori*, even specifying the values of whatever coefficients the model contains. This is a more difficult task, but more interesting.

It is my working hypothesis that populations of fringillids (sparrows) are limited by the availability of winter food. Then the population size of these species at the end of the winter is determined by the available winter food. Habitats in winter are, I think, of two sorts. There are the optimum species habitats in which the surviving individuals reside through the winter, and there are the fringe habitats that excess individuals reside in until death. The latter are of interest behaviorally but are of no importance in assessing the size of populations. Their winter densities are largely determined by the extent of overbreeding the previous summer. Their availability has little affect on densities in other habitats at any season. The former habitats, the prime species habitats, are important; their availability and richness are primary in determining species abundance at any season.

My intent in this chapter is to develop some primitive ideas about the regulation of fringillid densities in optimal wintering habitats. I shall develop a theory of the number of bird-winters of food in each habitat. I presume that the model constructed for wintering great tits (Chapter 3) can satisfactorily incorporate the deductions of such a theory. It will be remembered that, given sufficiently efficient dominance regulations and sufficiently prolific breeding, the surviving winter population was a very specific function of the available bird-winters of food and the inevitable mortality rate over the winter. I shall in the following discussion begin an attempt to deduce *a priori* the first of these variables. I hope eventually to see entire winter suitability (survival?)-density curves deduced *a priori*.

THEORY

The problem of "winter" densities in fringillids is an example of the general problem of the regulation of density by food when food is "limiting." There exists a certain amount of available food. An organism utilizing this food requires, for survival, a certain amount of it. Some fraction of the available food is utilized, so the number of organisms living on this resource is equal to the total amount taken, divided by the amount needed (taken) by each organism.

There are three components to this kind of analysis of density: the food requirements for maintenance of the organisms, the efficiency of utilization (including discovery and capture) of the available food, and the production of available food. If these three components are properly understood, K (bird-winters of food) can be explicitly deduced. Thus, this aspect of population ecology has its roots in physiological ecology, adaptive morphology, and the energy structure of ecosystems. We proceed in an attempt to relate these areas to the regulation of density in wintering fringillids.

FOOD REQUIREMENTS: A THEORY OF SPARROWS' BODY SIZE

Given a limited food supply of known quantity, we can state the amount of energy flowing into and through a species. In order to understand the population density of the species, however, we must consider whether this energy flows through a large number of small organisms or a small number of large ones. Thus, we need a theory of body size.

Presumably the body size of organisms is adapted to the selection pressures in the species niche. These pressures fall into three general categories: those concerning the effect of body size on metabolic requirments; on processing and digesting; and on the ability to find, catch, and defend resources.

Required Energy and Size

Generally, in homeotherms, metabolized energy increases with body size but the relationship is not linear. Instead, metabolism is proportional to body weight raised to a power less than one, usually about .7. It is not certain why this nonlinear relationship holds. There are several plausible possibilities, including changes in surface-volume relationships with size, changes in shape and composition with size (e.g., Thompson, 1917), and changes in activity and power needs with size. In a hummingbird-vulture comparison, for example, the vulture has less surface area per gram of weight, proportionally more nonmetabolizing bony and connective tissue, and in flight, frequently soars, using little or none of its own power. However, when the vulture must power itself, it must provide energy proportionally to its weight. The effect of all these factors average in some unknown way yielding the

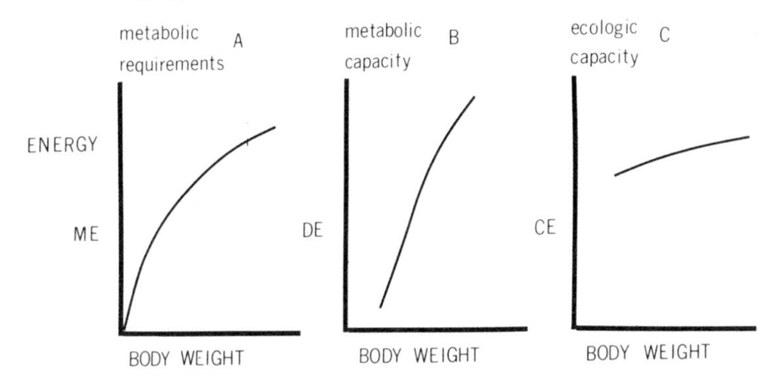

FIGURE 52. Principles in a theory of body size. A, B, and C are energy versus body weight graphs, metabolic needs (ME) are described as proportional to BW raised to the .7 power. Digestive capacity (DE) is more nearly proportional to BW, and food catching (CE) is some unspecified function, approximately horizontal in the regions of interest.

observed .7 exponent. Figure 52A provides a hypothetical graph of such a relationship.

Digestive or Metabolic Capacity and Size

The ability to digest and process food also should increase with size. There is less evidence pointing to one particular nonlinear relationship. Surface-volume relationships, undoubtedly important in digestion in vertebrates, can be ignored because of the coiling of the gut. If the gut is a tube and if increases in weight simply involve lengthening of the gut, the absorptive area should increase proportionally to weight. However, the supportive tissues of the body which house the gut will also have to increase in strength with an increase in cross-sectional area, resulting in an increase in weight in this tissue component as well. This weight increase will depend on the cross-sectional area raised to the $3/2$ power (Thompson, 1917), and so will be greater for larger increases in gut weight. As was true for metabolic requirements, this implies that digestive capacity will increase with body size raised to some power less than one. The degree of nonlinearity, however, might well be smaller in the case of digestive capacity, as both the linear and nonlinear effects are prominent. In the case of metabolic requirements, most of the theoretical arguments favored a reduced per gram energy requirement for larger animals. To test this possibility, some of the data reviewed by Kendeigh (1969) may be appropriate. He provides data (Table 8) both from his laboratory and elsewhere for temperate species, giving the *maximum rate of metabolism* (per bird per day). It is possible that this maximum (in winter-adapted birds at least) is limited by digestive rate. Increased photoperiod, which permits more time for feeding and digestion, sometimes results in an increase in the maximum (e.g., Zimmerman, 1965). Thus, we can estimate the exponent of the BW-DE relationship by considering the slope of a log-log regression.

Kendeigh's data are of mixed quality, and we shall consider several analyses. Using only the three temperate species for which exact figures are given, an exponent of 1.4 is obtained. Kendeigh also gives underestimates of

TABLE 8. Body weight and maximum food consumption

Species	Weight	Maximum metabolism (kcal/day)
Field sparrow	13.6	19.0
White-throated sparrow	27.4	32.9
Common redpoll	14.0	21.4
Hoary redpoll	15.0	24.3
Tree sparrow	19.0	32.8
House sparrow	25.2	>35.8

After Kendeigh (1969).

DE for two temperate large species, the house sparrow and the evening grosbeak. One of these is clearly less biased than the other (see Table 8). The house sparrow value was obtained at −31°C (−23°F), and this species probably cannot tolerate much colder temperatures. However, an evening grosbeak value was obtained at −20°C (−4°F). This is a boreal species and must certainly be able to tolerate much lower temperatures. Therefore, including just the house sparrow figure in the analysis (which still must yield an underestimate of the slope, as the house sparrow is larger than average sparrows), an estimated exponent of .9 is obtained. Kendeigh also provides some figures for Gulf State winter residents (Table 8—field and white-throated sparrows). When these are included in the analysis, an exponent of .8 is obtained.

These exponents all are greater than the .7 value obtained for metabolism, and there is some reason to believe that two of the three are somewhat underestimated. The data are rather sparse, however, and more stimulating than conclusive. Without claiming correctness, the present theory will assume that digestive capacity increases nearly linearly with body size as in Figure 52B.

Finding, Capturing, and Defending Resources

The ability to find, capture, and defend resources may or may not depend on body size. For species that feed on seeds in a temperate winter, the superabundance of seeds most of the winter season precludes much of a finding problem, and normal variation in body weight should not have any effects. The same is true of capturing a seed, so defense alone can be a significant factor in the utilization of available environmental resources. Size can influence dominance (see Figure 40), so the ability to defend resources should increase with size.

The rates of finding, capturing, and defending deter-

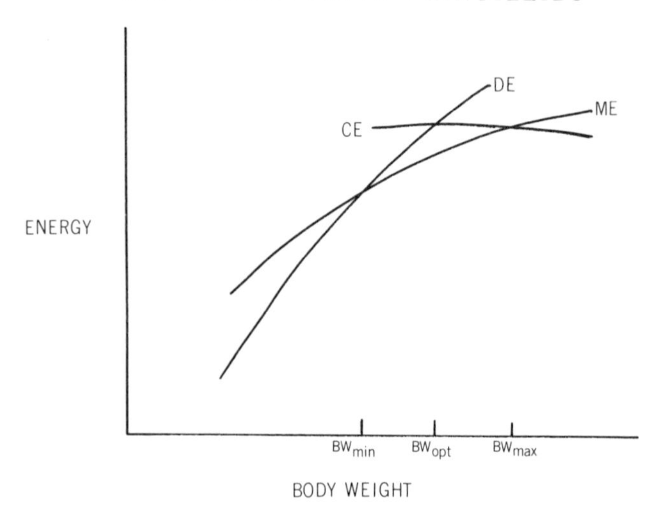

FIGURE 53. Theory of body size. Figures 52 A, B, and C are combined here, to yield minimum, maximum, and optimum body sizes.

mine a potential rate of consumption of seeds. This rate may increase slightly with size, depending on how closely size is correlated with dominance and on how important (dominance) defense is in rate of consumption. My observations on the junco indicate that dominance largely affects *where* an organism feeds, and need reduce the actual rate of consumption only slightly.

In ignorance and for simplicity, I shall assume that the consumption-rate-body size curve looks about like that in Figure 52C.

The three kinds of curves in Figures 52 A, B, and C are all energy versus body weight graphs and can easily be plotted on one graph. This is done in Figure 53. In this figure, a minimum body size is specified, below which the animal cannot digest fast enough to compensate for heat losses. Also, a maximum body weight is specified, above which the animal cannot find food fast enough to support its energetic needs. Organisms with sizes between these extremes can find and digest more than enough resources to meet needs. There is also specified an energetic "optimum" body size, where the organism can catch only as much as it can digest or where it can digest all that it can catch. At this point, the organism has the greatest absolute "productive" potential, given the above assumptions on the curves.

There are thus three body sizes of interest; a minimum, which represents the peak of overwinter survival efficiency (least amount of food required); a maximum, which represents the minimum permitted overwinter survival efficiency; and an optimum, which is the "safest" efficiency. In the following section we shall try to decide which (if any) of the body sizes will be truly adaptive.

Varying the Consumption Curve

The rate at which an organism can find food depends on the density of food in the environment, the amount of time spent looking for food and the organism's efficiency of feeding. For the particular organism described in the above theory, we have presumed that the feeding efficiency is nearly fixed. However, both the density of food and the proportion of time spent feeding might be influenced by competition. Under conditions of high-population density, we would expect that the average or integral of density of food over the whole season would be lower, and that the organism would be required to spend more time in territorial defense or other forms of social interaction. Then, under conditions of competition, the consumption curve would be lower. This would lower the maximum possible body size, forcing it toward the minimum determined by the digestive curve. Of course, competition alone cannot make it fall below this minimum.

This leads us into an interesting argument. If a species is food-limited, competition should force it toward (maybe even to) the minimum body size. If it is not food-limited, it must be limited by other factors. Suppose it is limited by predation or weather. Then, for wintering fringillids at least, there is considerable advantage to having a productive potential that allows the organism to make up losses sustained in escaping (or hiding from) a predator or in waiting out a storm. Such a species might tend toward the "middle" optimum. This productive optimum need not be close to the lower or upper limits of body size, as competition has not forced the extremes together.

It is also possible that a species is limited by breeding resources and in the breeding season is exposed to selection for large size. Then the body size should tend toward the maximum in the above theory. For example, some icterid males are strongly territorial in the breeding season where available habitats may be limiting. Large size, for males at least, is an evident advantage in territory defense, so we expect the males (but not the females) to approach the maximum possible wintering size. Of course, if there is somehow selection for smallness in breeding birds, then the digestive minimum will be approached.

In summary, if winter food is limiting, body size should tend to the minimum; if predation or weather is limiting, body size should tend to be at a productive optimum, which may be quite different from the digestive minimum. If the breeding habitat is limited and if breeding territorial activity is selecting for large birds, the body size of the territorial sex should tend to the maximum allowed by the rate of consumption.

Body Size as a Function of Food and Environment

In the populations of wintering sparrows that I have been concerned with, competition for winter food seems to me to be the most plausible of the three alternatives above. I now want to consider the effects of variations in

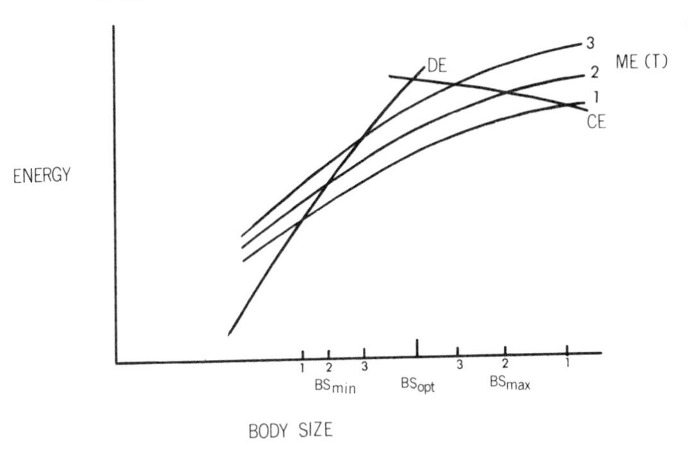

FIGURE 54. Theory of body size, where only environmental temperature varies. Three ME curves are drawn for three environmental temperatures. The uppermost curve (3) is for coldest temperatures (northern) climates, the lowermost for warm (southern) climates. The BS_{min} gets larger in colder environments, the BS_{max} smaller, but the BS_{opt} does not change.

temperature and seed size on optimum body size of such winter-limited sparrows.

There are several kinds of effects which we must consider for sparrows: these include the size and digestibility of seeds, the photoperiods, and the ambient temperature. Metabolic energy, ME, depends on the last of these and digestive energy, DE, on the first. In Figure 54, the theory is drawn for a series of habitats of varying temperatures, all of which, I assume, are below thermoneutrality. The ME curve is higher at colder environmental temperatures as the body temperature is constant and the rate of heat loss increases with the difference in temperature between the body and the environment. A linear plot of DE versus body size is also drawn. It shows an increase of the minimum body size ($BS_{min} \approx BS_{opt}$) for organisms eating the same size seed in different areas of similar photoperiod

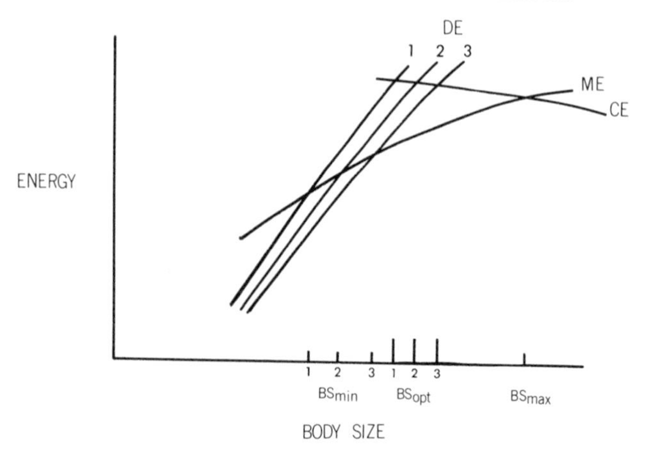

FIGURE 55. Theory of body size, where only food digestibility varies. Curve 1 represents the most digestible fare, and 3, the least. The minimum and optimum body sizes increase as food becomes less easily processable, but the maximum body size does not change.

with decreasing temperature. Since $BS_{min} \approx BS_{opt}$ for winter food-limited species, Figure 54 predicts Bergmann's rule for them. In Figure 55, one ME curve is drawn with a series of DE curves. These DE curves may represent (for a seed-eating bird) seeds of varying size, as noted. The bigger seeds take longer to digest and so require a bigger body size for organisms living in the same area. This figure can also represent changes in DE with varying photoperiod, as less energy can be digested per day if there are fewer daylight hours. Thus, if temperature and seed size are the same, an increasing photoperiod will permit a smaller BS_{min}.

Thus, this theory specifies the optimum body size for a sparrow consuming a certain size of seed, in a habitat of some photoperiod and temperature. However, some, probably all, of these factors vary markedly and a single individual bird is exposed to many photoperiods, many temperatures, and even many seed sizes. Let us presume a conservative strategy. A bird overwintering in a given habitat should be adapted to the largest seed size consumed on the coldest day of the shortest photoperiod. This conservative strategy is plausible; it does a bird no good to survive very efficiently for 99 percent of the winter days if it is likely to die on the remaining 1 percent. It also gives us very particular values for the variables for each bird in each habitat. Newton (1967) offers data to support the notion that only on the most stressful days of winter are birds forced to eat their specific diet.

TESTING THE THEORY

The theory makes several predictions. Body size of sparrows should increase with decreasing temperature and with decreasing photoperiod. It should increase with seed size. Thus, this theory has the necessary virtue of being testable. ME-BW curves at varying temperatures are already available for fringillids (Helms, 1968; Figure 56).

I have attempted to measure digestion-rate curves for different-sized birds and different-sized seeds, using whole natural seeds of different sizes (mostly vegetable and clover seeds). Unfortunately, the sparrows would not eat these seeds. In fact, most of them would eat only millet; some also ate sorghum. Difficulties were also encountered in getting the birds to eat at maximal rates, although presumably this would not be difficult in a cold-environment chamber; Kendeigh has successfully used such a chamber with artificial foods. These values need to be estimated for foods as similar in structure and texture to natural seeds as the birds will eat.

Given laboratory estimates of the ME-DE curves in this theory, the values of BS_{min} for any seed size, photoperiod, and environmental temperature can be estimated. BS_{min}

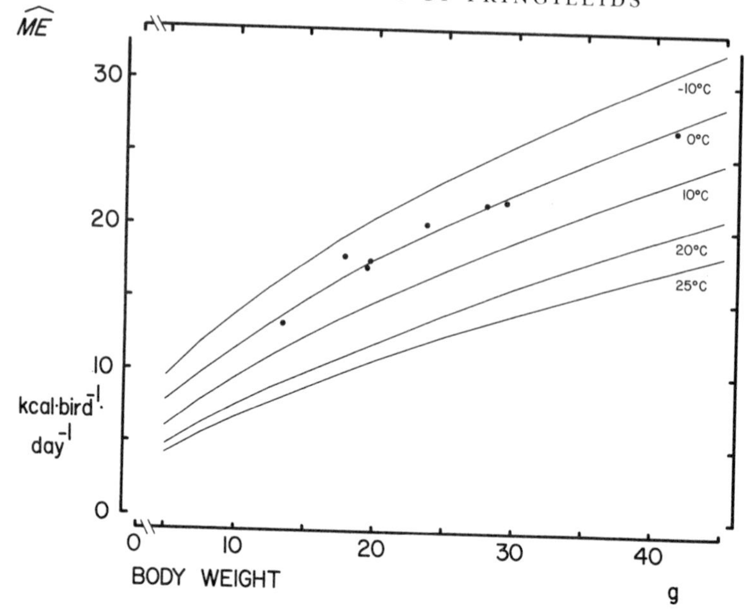

FIGURE 56. Estimated metabolizable energy (ME) for winter-acclimatized birds in relation to body weight at several temperatures. Calculated data points are given around the 0° C line for eight species of buntings. (After Helms, 1969.)

values can then be predicted from analyses of seed sizes in stomachs of wintering fringillids collected on the morning of the coldest days of winter and the actual body weight compared with the predicted value.

In trying to assess the assumption that $BS_{opt} \approx BS_{min}$ – that wintering sparrows are limited by the rate at which food can be processed – I have made some observations which are relevant. While watching populations of field sparrows that were neither baited nor netted, two kinds of movements were noticeable. These birds seemed to feed in one area (on broomsedge, *Andropogon virginicaus*) in the morning and in other areas (on crab grass, *Digitaria*) in the afternoon. And when the weather got colder, the populations all moved into broomsedge habitat. *Digitaria* seeds are the typical diet of the larger savannah sparrow and seem somewhat larger than broomsedge seeds. Broomsedge is a more typical diet of the field sparrow. When savannah sparrows were abundant, field sparrows were almost never seen eating *Digitaria.* Thus, the above observations suggest that field sparrows eat atypically large seeds on warmer days or in the evening. This suggestion is reasonable, given limitation by rate of digestion. On warmer days, the ME curve is lower and the body size of field sparrows is then appropriate to a larger (*Digitaria*) seed (Figure 57). Also, a field sparrow going to roost has all night to digest what is in its crop and need not choose the presumably more quickly digestible broomsedge seeds then. Only in the early part of the day can there be any

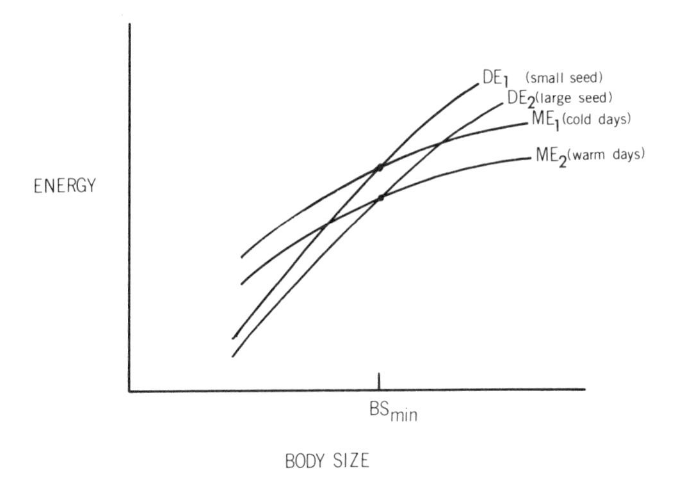

FIGURE 57. Changes in diet with environmental temperature. A food-limited sparrow near BS_{min} in body size will on cold days (1) have the body size appropriate to a small seed but will on warm days (2) have the body size appropriate to a larger seed. Such a bird might change diet accordingly, if the two kinds of seeds are equally available.

limitation of rate of digestion, when the emptying gut can be refilled.

These observations need to be extensively confirmed; they are presented for the predictions they suggest. We may predict that, given a choice of diet, larger items will be selected on warmer days and in the evening.

I made a preliminary test of the latter of these predictions. I placed partially ground and sifted millet (two sizes, cracked and whole) on two feeding trays in the field. Wild field sparrows were permitted to feed at these trays for ten minutes in the morning and ten minutes in the evening. The average number of birds eating at each tray over the ten-minute period was estimated by averaging over counts made every fifteen seconds during the period. The final measure used was the proportion of activity at the tray of cracked millet. The trays were shifted in position at random (coin toss) at first, but the preference for the cracked size was so great that I moved it out into a more exposed position until the two foods were used more equally. Then morning-evening comparisons were made. The results are in Table 9. There was a clear morning-to-evening shift in preference towards the larger seeds, as predicted.

Table 9. Food selection (percent choices for small *)

Date	Jan. 31	Feb. 1	Feb. 2	Feb. 3	Feb. 4	Mean
Field sparrow						
Morning	66	67	93	63	62	70.2
Evening	39	59	59	53	49	51.8
						$t = 2.7$†
Junco						
Morning	28	40	–	46	35	37.2
Evening	50	65	–	42	44	50.2
						$t = 2.0$ NS†

* Computed by taking the average number of birds at the tray of cracked millet, divided by the average number of birds at both trays.
† $P < .05$. NS = no significant difference.

The same study was simultaneously conducted on the junco, which is a larger bird. The prediction should not hold, of course. If millet is of appropriate size for the field sparrow, then the larger junco presumably has a potentially large positive energy balance even on the whole millet and can afford to eat it even in the morning. The results are less clear, although they suggest a shift in preference toward the larger seeds in the morning. Such a shift could be due to competition with the field sparrows, which avoid the larger seeds in the morning, leaving these available to the junco. I did not extend the study into spring to observe the effect of temperature on seed-size preference, but I predict a general shift to larger seeds.

A further application of the theory is interesting. A general correlation between seed size and body size is predicted for fringillids. This correlation leads to the prediction that sparrows feeding in habitats where larger

TABLE 10. Winter habitat and body size of fringillids

Forest (deciduous and pine)		Shrub		Open field	
Species	Weight (gm)	Species	Weight (gm)	Species	Weight (gm)
Evening grosbeak	55				
Cardinal	45	Cardinal	45		
Towhee	40				
Fox sparrow	35				
White-throated Sparrow	30	White-throated Sparrow	30		
Purple finch	28				
		Song sparrow	21	Song sparrow	21
		Junco	19	Junco	19
				Savannah sparrow	19
		Field sparrow	12	Field sparrow	12
Pine siskin	11	Pine siskin	11		
Goldfinch	11	Goldfinch	11	Goldfinch	11
Average	31.9	Average	23.6	Average	16.6

Analysis using the model: $BS = B_0 + B_1 H$ where $H = -1$ for open field, $= 0$ for shrub, $= +1$ for forest was statistically significant; $r = .4447$ (19 d.f. $p < .05$).

seeds are found will be larger. There is, however, also a positive correlation between seed size and plant size. It is not perfect, but it nonetheless exists (Salisbury, 1942). Then we predict that habitats with larger plants should be inhabited by larger fringillids. Table 10 lists some habitats found around Raleigh, North Carolina, in decreasing order of plant size. With each habitat are listed the fringillids which are commonly found there, with their body weights. The larger species are found in forest and shrub habitats, confirming this prediction.

UTILIZATION

Distribution of Resources in Kind

Given well established curves for ME and DE, and supposing that the optimal body size is close to the minimum body size (or some consistent function of that minimum), we could presumably go into a field or forest, sample the seeds that birds would be likely to feed on, and from the sizes of the seeds and the lowest temperatures in the winter in the habitat, deduce what size or sizes of seed-eating birds should be present. Knowing how many seeds were present and how many could be found and utilized would tell how many calories were available to the birds. From the body sizes and temperature data, we could estimate how many calories each bird would need. Dividing the per bird-per winter requirements into the calories available in the habitat would let us predict the density for each size class presumed to be present. However, if we actually undertook this study, we would quickly encounter a problem.

The seed study would almost certainly yield a continuous size distribution of available seeds. The theory tells us that if we are given point values of seed sizes available in certain quantities, then we can predict the body size and number of sparrows which will be present. But we shall not be given fixed seed sizes and numbers. We shall have instead a distribution that ranges continuously over some range of values. What then, will determine the distribution of organism sizes feeding on these seeds?

In essence, there will be very few seeds of any given size, so the bird of body size appropriate to some seed size will not be able to find enough of these to exist. Instead, it will have to consume a range of seed sizes. Presumably, the bird will choose seeds as close as possible in size to its optimum. In fact, the organism should choose some distribution of seed sizes which would at least include its optimum. This distribution would of necessity reflect the availability distribution, but would always diminish away from an organism's optimum.

The organisms feeding on these seeds, as a group, presumably eat all that are available because this resource is assumed to be limiting. So we know that the individual utilization distributions must sum up to the total availability distribution, if all distributions are expressed as frequencies multiplied by the total taken or available.

One possible situation (fine-grained, MacArthur and Levins, 1964) would have all organisms at some intermediate body size. The utilization distribution for each organism would then be expressed in frequencies, equivalent to the availability distribution for all wintering sparrows. This situation would probably be quite unstable as competition would be quite severe at the seed size which is optimal for the organism size, and less severe elsewhere. An organism somewhat larger or smaller would in this case presumably have a selective advantage.

Another solution, also extreme (coarse-grained), would have the organisms distributed in size as the seeds. Each organism would specialize as much as possible, taking *all* seeds that he needed for survival from a range about his optimum. This case also has a difficulty: Each organism is restricted to a very particular set of seeds in the environment and cannot be presumed to be able to travel sufficiently fast to find enough of them (MacArthur and Pianka, 1966).

A third solution would have the range of utilization of each organism fixed (actually, we fix the variance of the utilization curve and its shape), and to pack as many types of organisms into the environment as possible. A simple but ingenious mathematical structure for this kind of solution has been developed by MacArthur and Levins (1967). Basically, if one organism type is of a particular size, it will most effectively utilize the resources at its optimum, but less effectively take the resources at some distance from its optimum. Another organism of a different size, but similar, also utilizes efficiently only its optimal resources. Presumably, the seeds of optimal size are now fully used; however, consider the seeds of size exactly intermediate to the optima for these two organism types. Seeds of this size are less than fully used by the two types separately but somewhat used by both. If, in fact, both use exactly half of these seeds, then all of these seeds will be used also. This implies that the organisms are a very particular distance apart, depending, as MacArthur and Levins have shown, on the variance of the utilization curves.

This last solution is a less extreme case of the first and vulnerable to the same criticism. The within-species (organism type) competition is concentrated at a certain seed size point. It would seem that variants which are optimally adapted to some different seed sizes should have some advantage. MacArthur and Levins proved, however, that this need not be the case. If nearby species (types) are similar enough, each *tries* to consume more than half of the intermediate (not optimum) sized seeds which are therefore effectively overutilized. An intermediate organism type receives enough interspecies competition to prevent successful establishment. So this solution, at least, remains a real possibility.

Distribution of Resources in Space

Before going on with the analysis of the effect of variation in resource quality on consumers and consumption, we must consider in some detail the problem of finding or capturing the resource. The seeds of wintering sparrows are, in general, located in one of three places: on the plant, on the ground, or under the litter, if any. There is also, in general, a tendency for different-sized seeds to be in different locations, within a single habitat type at least. In order to take a certain size seed in some particular area, therefore, a bird must generally be perching upon the plants or walking on the ground or scratching through ground litter. Each of these activities requires a different kind of leg. For example, in Figure 58, relative length of tarsus is taken as the measure of leg adaptation. This measure is compared to a perching index, which is a rank ordering of different species as to frequency of perching, and a scratching index, similarly obtained (Fretwell, 1969b, provides details). It can be supposed that every kind of feeding niche has associated with it a particular length of tarsus. Thus, feeding locations can be reduced to a single adaptive dimension (number) represented here by tarsus length. This dimension possibly reflects perch stability (Grant, 1966), defined as the amount of time the bird's center of gravity is over its feet.

Morph Diversity

We can now construct a three-dimensional representation of the niche. One dimension will be the seed size, one dimension will be the feeding location (perch stability),

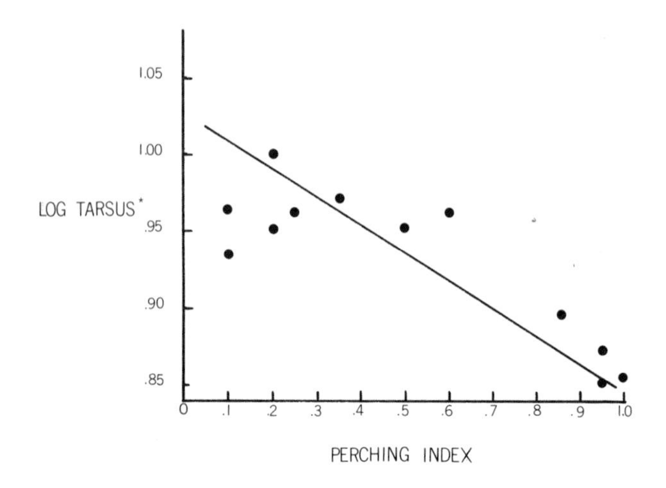

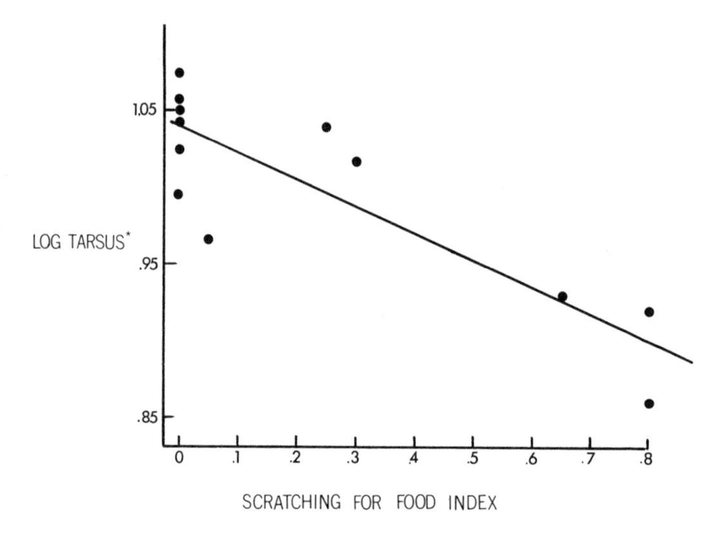

FIGURE 58. Log_{10} tarsus versus trophic locomotor activity in some fringillids. Tarsus length decreases with more time spent perching while feeding or spent scratching for food. *Log_{10} tarsus is corrected for body weight and the other trophic feeding behavior (i.e., for scratching, if perching is being considered). (From Fretwell, 1969b.)

and one dimension will be the abundance of food (bird-winters, K) supplied by the niche. (Dimensions correspond to resource quality, resource location, and resource abundance.) In effect, we have a two-dimensional frequency distribution, a surface with hills in regions where there are many seeds of a certain size and located in places where perches are of a certain stability. There are valleys elsewhere. We have taken the original seed size and frequency distribution and spread it out over a new dimension.

This process, which we shall presume really exists in nature, is analogous to two-way paper chromatography which is done to separate different species of molecules from each other. If the analogy holds, the effect should be the same: The multidimensionality of niches should tend to separate kinds of resources, and therefore species of organisms. This moves away from the continuous, uniform resource distribution assumed in MacArthur and Levins' (1967) theory of limiting similarity. As noted above, this theory proved that if a species was closely enough bounded by other species feeding on a resource *uniformly distributed along some relevant dimension,* there could be no invasion and only a certain number of optimal morphs could coexist. In effect, competing species overutilized the resources between their respective optima, depressing the carrying capacity in this intermediate region below zero even for the species (morphs) better adapted to these intermediate optima.

But suppose the resources and species are separated and do not grade smoothly into each other; that, in fact, the species are held far from the limiting similarity because the resources of one species do not grade smoothly into the resources of another. Then the theoretical region of interest in resource distributions is the outer boundaries of the resource-abundance curves, not the resources "between" species. Species are then limited in abundance by the availability of resources and not by competition. Each has its own patch or kind of resources, on which it is very efficient, and on which all other species are very inefficient.

The question to be asked then is as follows: Suppose there exists a certain "mound" of resources, distributed unimodally over a small range of seed size and perch types. Further, suppose that only a single morph is present which is optimally adapted to the most abundant resources available. Is this system stable or is it subject to invasion by other morphs? This is a typical two-species competition problem, which is also solved by MacArthur and Levins from the population growth differential equations:

$$\frac{dN_1}{dt} = \frac{rN_1}{K_1}(K_1 - N_1 - \alpha_1 N_2)$$

$$\frac{dN_2}{dt} = \frac{rN_2}{K_2}(K_2 - N_2 - \alpha_2 N_1)$$

where N_1 = resident species, N_2 = invader morphs.

Since we are interested in the outer boundary of the resource curves, we may presume that K is steadily decreasing as the resource quality and location move away from the optimum for the resident species. That is, the resident species (morph) presumably has the highest value of K possible for any single morph living alone on the resource. Any other morph (body size–tarsus length combination) is optimally adapted to some resource kind that is rarer, or some resource location that has less resource in it. Then its K is lower. So we assume that $K_2 < K_1$.

By MacArthur and Levin's argument, species 2 cannot invade unless $dN_2/dt > 0$, that is,

$$(K_2 - N_2 - \alpha_2 N_1) > 0.$$

But $N_2 \approx 0$ (by definition of invasion) and $N_1 \approx K_1$, so we have

$$K_2 - \alpha_2 K_1 > 0,\ \alpha_2 < \frac{K_2}{K_1}. \qquad (20)$$

What determines α_2? This is the competition coefficient representing the efficiency of the resident organism in the invading organism's niche, relative to the invading organism. We can be more specific in the present case of an invader-resident system. We can, in fact, consider a small region around the invader's optimum on the seed size-perch stability plane, and can consider what resources in that region the average resident utilizes when only the resident exists in the system. This quantity clearly depends on the breadth of the resident's utilization curve. If it is very broad the resident consumes almost as much (as a percentage) of this resource as of its optimum resource. If the resident is relatively specialized, it consumes relatively little of its suboptimal resource compared to its consumption of its optimal resource. We shall estimate α_2 as the ratio of suboptimal resources consumed by the resident 1 to optimal resources consumed, *if* all resources are equally abundant. Therefore, $\alpha_2 K_1$ is proportional to the maximal amount of resource 2 possibly consumed. If this is less than K_2, then some resources are left unconsumed to support invasion by species 2, so invasion occurs when $\alpha_2 K_1 < K_2$, which is identical to Inequality (20).

This rather easy result allows us to make some predictions. Clearly, an environment which encourages specialization (i.e., a coarse-grained environment) invites the evolution or invasion of a range of morphs, while an environment that must be utilized in a fine-grained way (broad utilization efficiency) is not so open (see Figures 59 and 60). Thus, we expect specialist species to evolve higher variances and more polymorphism.

By way of testing this prediction, we note the following. Some potentially specialist bird species are: accipiters,

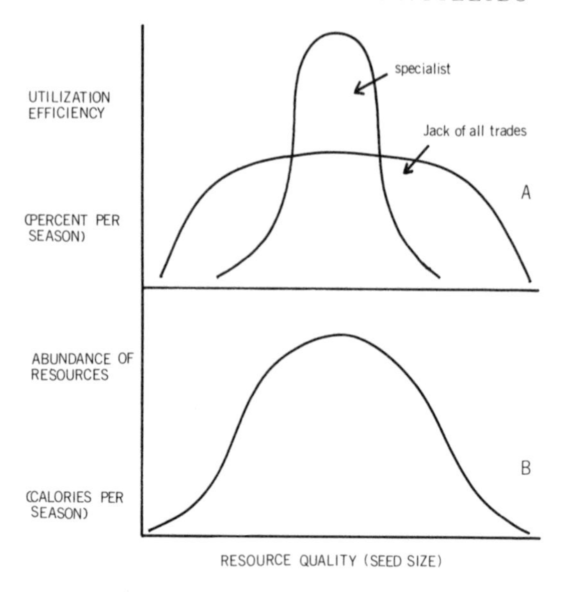

FIGURE 59. Abundance and utilization efficiency of resources. In A, the utilization efficiency is described for a specialist and a Jack of all trades. In B the available resources are given.

which are hawks that pick and pursue; woodpeckers, which can select branch sizes; fringillids and other seed eaters, the resources of which are easy to identify and catch separately. We can therefore expect much variation in body size (e.g., sexual polymorphism) in these species. Generalist (nonselective) species, however (swallows, swifts), typically should lack extensive variation or sexual polymorphism (see Figure 61). That such sexual polymorphism permits broader niche utilization is well documented (Selander, 1966).

Another test of this prediction can be made. Emlen (1966) and MacArthur and Pianka (1966) have shown that an organism feeding on a sparse resource cannot afford to bypass much of what is found, and so must have a fine-grained pattern of utilization. Such a Jack of all trades should have a low but broad efficiency curve. Thus each morph type is likely to take up more of the island of resources. Then fewer morph types can coexist and the variation of the species should be lower. Rare resources are frequently of two types: (1) those that are on the tails of a distribution, and (2) those resources on a higher trophic level. We consider type (1) first.

The smallest and largest sizes of seeds should be on the tails of the overall available seed-size distribution. Sparrows which feed on the smallest or the largest seeds should have the lowest variability. From the body-size theory, we know that these birds should be, respectively, the smallest and the largest in the area; we therefore expect the varia-

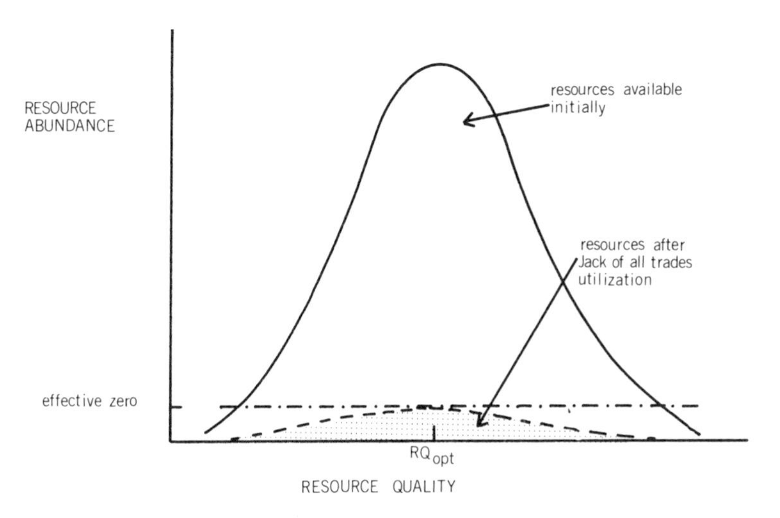

FIGURE 60A. Effect of Jack of all trades utilization on resource abundance. The generalist's population is limited by the resources to which it is best adapted (RQ_{opt}), and these resources are depressed to an effective zero level at which the generalist can no longer increase its population size. Because the generalist utilizes other, rarer resources about as efficiently as it uses RQ_{opt}, these are depressed *below* effective zero, and no other form can exist on the resources. Note that a constant value of effective zero assumes all morphs are equally efficient on their optimal resource.

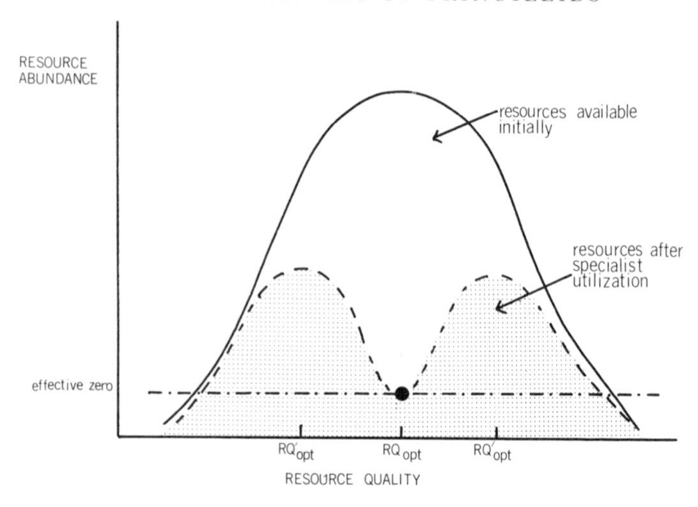

FIGURE 60B. Effect of specialist utilization on resource abundance. The specialist population is limited by the resources to which it is best adapted (RQ_{opt}), and these resources are depressed to the effective zero level at which no consumer can further increase its population size. Because the specialist does not utilize rarer resources as efficiently as it uses RQ_{opt}, these remain at levels above effective zero, and can be utilized by other organisms adapted to different kinds of resources (RQ_{opt}). Such a species may generate ecotypic variation to produce morphs which can utilize these different kinds of resources.

bility to increase with average body size up to a maximum (where resources are most abundant) and then to decline. We observe in Quay's study (1947) that the field sparrow and the savannah sparrow are the most abundant fringillids wintering in the Piedmont Carolina area. We expect these species to have the highest variability. Birds smaller and larger than these species should have lower variability.

Figure 62 gives an analysis of data collected by me which shows a significant increase in the coefficient of variation (c.v.) from the smallest fringillid, the goldfinch, to the field sparrow, and a significant decrease thereafter to the very large evening grosbeak. This confirms the prediction of the theory.

Note that the increase from the goldfinch to the field sparrow is very rapid while the following decline is gradual. The goldfinch, while smaller than the field sparrow, is only slightly smaller. This confirms part of the conservative assumptions made earlier: Sparrows would be adapted to the largest seeds in their range of utilization. The goldfinch

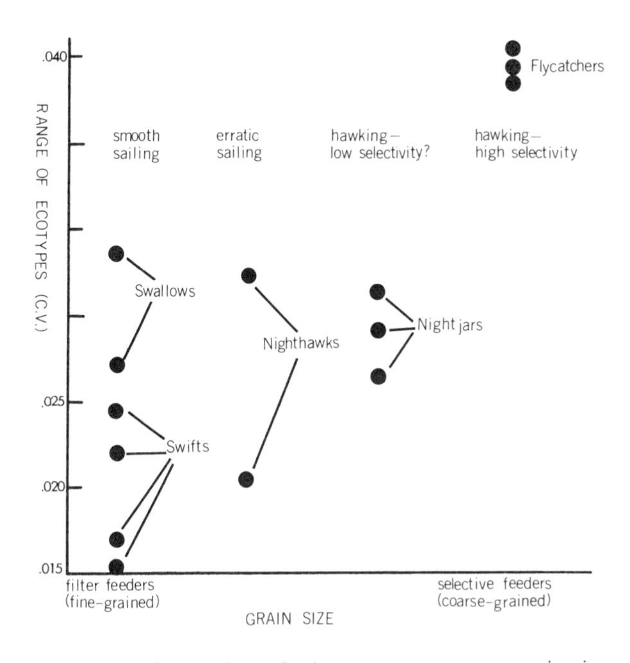

FIGURE 61. Aerial plankton feeders; ecotypes versus grain size. The theory of Figure 60 predicts that there will be less species variation in fine-grained feeders which are Jack of all trades types than in coarse-grained (specialist) feeders. Plotted is the coefficient of variation of wing length in museum specimens versus selectivity of feeding. Presumably, swifts and swallows which are "smooth-sailing" do not go much out of their way for any given insect. Thus, the *extra* cost per insect is low and the birds can afford to be less selective (MacArthur and Pianka, 1966). Flycatchers, which foray out from a sitting place for each insect, spend considerably more *extra* work per insect, and can thus afford only insects of a particular (high) energy return. Specimens were from the Kansas Museum of Natural History.

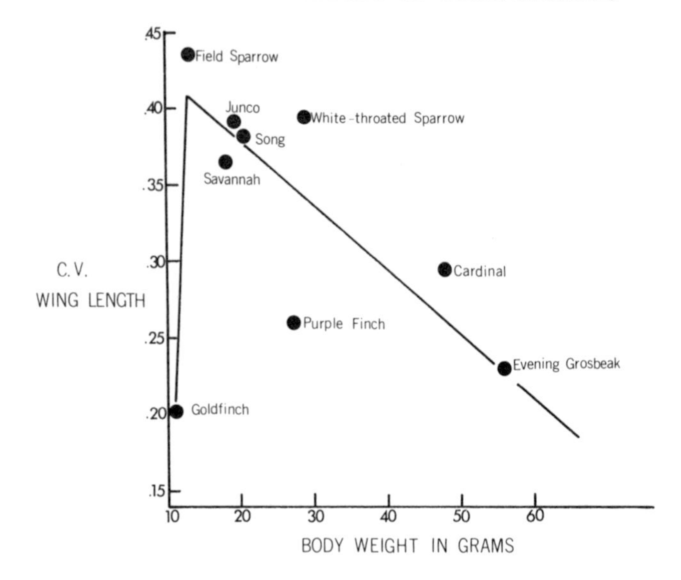

FIGURE 62. Body size variation and resource abundance. Assuming that intermediate-sized resources are most abundant, the intermediate-sized birds can afford to be selective in the feeding. The limiting energetic process becomes processing the seed once found, not finding it, and so selectivity is advantageous. Selective individuals in a species leave some of the species' resource uneaten (Figure 60), and so make it advantageous for the species to generate ecotypes, that is, to become more variable. The line trends are statistically significant. Note further that species above the line (field sparrow, junco, song and white-throated sparrow, and cardinal) tend to be sedentary in winter, forming stable winter populations. Those below or on the line (savannah sparrow, purple finch, and, perhaps, evening grosbeak) tend to be erratic to engage in extensive midwinter movements. Sedentary behavior permits greater resource selectivity, as the theory predicts.

must be about as large as the smallest field sparrow, since the largest seeds taken by a goldfinch will be about as large as those taken by a small, very selective field sparrow. However, large birds, which must also take a wide range of seed sizes and which are also adapted to the largest seeds that are normally taken, have to be considerably bigger than the largest of the next smallest species (see Figure 63).

The second type of presumably fine-grained consumer feeds on a resource that is rare because it is on a higher trophic position. For example, insectivorous birds at Raleigh are considerably rarer than graminivorous birds (Figure 51); we therefore expect them to have a lower variability. My own data are meager, but Table 11 clearly supports this prediction as well.

Now we can speak somewhat more usefully about the

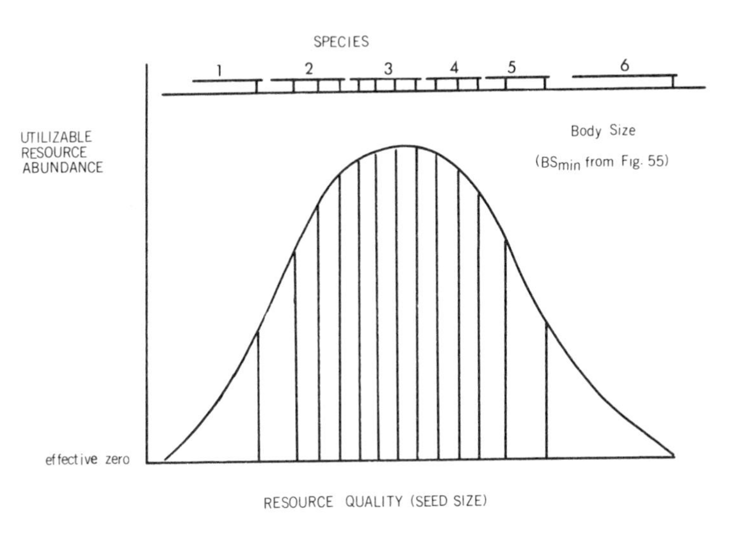

FIGURE 63. Utilization of resources by individuals and species. The total curve represents all resources available each day. The separate divisions enclose equal areas, and represent the similar caloric needs of organisms living on different kinds of resources. Thus, the organism type living on the smallest seed size must utilize a wider range of seed sizes, because small seeds are relatively rare. Organisms living on middle-sized seeds can afford to be specialist and to have more ecotypes. On the upper abscissa, body sizes corresponding to various seed sizes are given and it is assumed that each organism is adapted to the maximum seed in its diet (each vertical line is the seed size for one ecotype). There are six species and each is allowed about the same range of diet types (horizontal line). The smallest and largest (1 and 6) are least variable (only one vertical line); species (1) is nearly as large as species 2, but 6 is much larger than 5. See Figure 62.

TABLE 11. Trophic status and species variance

	Fall migrants * d.f.	c.v.(SD/$\bar{x}$)
Vegetarian species		
Towhee	5	.038
White-throated sparrow	8	.047
Miscellaneous Fringillidae †	8	.037
Weighted mean		$\bar{x}$ = .0410
Mixed diet species		
Catbird	33	.031
Brown thrasher	11	.037
Swainson's thrush	29	.033
Myrtle warbler	41	.034
Weighted mean		$\bar{x}$ = .0332
Insectivorous		
Yellowthroat	54	.034
Magnolia warbler	14	.036
Black-throated blue warbler	11	.026
Brown creeper	20	.030
Redstart	37	.034
Ruby-crowned kinglet	12	.032
Weighted mean		$\bar{x}$ = .0328

* Data from R. Teulings, P. W. Sykes, Jr., and myself, collected during Operation Recovery work on North Carolina's outer banks, fall 1966.

† Includes estimates of c.v. from 2 swamp sparrows, 2 juncos, 2 field sparrows, 3 field sparrows (diff. sample), 3 cardinals, and 2 towhees.

kinds of organisms using resources distributed in quality. Due to the separation by length of tarsus, we have, perhaps, islands of resources (which we could call niches). Feeding on each "island" are particular organisms, if it is big enough to support a population. If the island is "small" relative to the species' range of utilization, all the organisms will be of a single kind and we can predict their abundance. If the island is "large," the organisms are of at least two different kinds.

We can make some vague statements about this situation. For example, the distribution of organism kinds will not be so broad as the distribution of resources, but it will be flatter. The first observation comes from the fact that organisms adapted to seeds (resources) near the edge of the seed size, location distribution (i.e., adapted to seeds that are similar to the rarer seed kinds) will perforce overexploit the very rare seed kinds near optimum, making it impossible for an organism well adapted to these rare seeds to invade. The second observation comes from the fact that overlaps occur between organisms in the utilization of seeds intermediate to the organisms' optima. Thus, seeds of a median character are vulnerable to attack from many sides, and will therefore be so sparse that fewer organisms adapted especially to them can survive. However, these average-type seeds are, presumably, more abundant originally. The increased utilization might simply cancel the effect of this increased availability.

If the islands are very large with plateau tops, the theory of MacArthur and Levins holds directly. Specialized monomorphic populations can appear, exploiting both intermediate and edge regions sufficiently to prevent other invasions.

Ecotypic Variation

The case of the large "island" of resources where the system will stabilize only with a broad range of morph types depends on the existence of ecotypic variation in the organisms. Thus, each organism must be able to specialize to some degree and select only its own niche for residence or utilization. Such ecotypic variation has been observed in the field, song, and savannah sparrows with respect to tarsus length and perching (Table 12), and in the field and

TABLE 12. Habitat variation in mean tarsus lengths (mm)

	Habitats							
	Most perching				Least perching			
	1		2		3		4	
Species	n	$\bar{x}$	n	$\bar{x}$	n	$\bar{x}$	n	$\bar{x}$
Savannah sparrow	5	22.04	10	24.18	37	24.31	28	24.30
Song sparrow	21	24.35	26	25.32			36	25.55
Field sparrow	40	21.00					40	21.43

In all three species, differences between habitats in mean tarsus length were statistically significant (Fretwell, 1969b).

savannah sparrows with respect to seed size and body size (Table 13). In the case of the field sparrow, some of the measurements of body size were made on breeding populations, although the evident selective advantage occurs in winter. This represents positive assortative mating which increases species variance, presumably in response to the selective advantage which we have shown that this variance offers. When survival versus wing length is plotted (Figure 64) a flat curve is found, consistent with the expectation that more than one morph type is optimal (frequency-dependent selection, Thoday, 1959). Thus, the ecotypic variation needed for reaching the variability predicted by the theory seems to be at least partly realizable.

One thing remains in this analysis: to establish the exact relationship between the distribution of morphs and the distribution of resources given utilization curves for the morphs. This problem has been partly solved for large "island" resources by MacArthur and Levins' theory, and for systems inhabited by generalist species or those on small "islands." Unfortunately, seed-eating fringillids fall

TABLE 13A. Savannah sparrow wing lengths and seed sizes in stomach

Section of plot 9	A	C	B	D
Number of birds (n)	9	3	4	12
Average percent *Digitaria* in gut (a small seed)	78.1 ± 9.55	65.33 ± 17.75	60.75 ± 18.99	57.58 ± 10.07
Average wing length, $\bar{x} \pm 1$ Standard Error (SE)	69.89 ± 1.18	70.67 ± 2.03	71.75 ± 1.60	71.33 ± .51
		Wing lengths		
	Number of birds	Mean	*SE*	
With sorghum present (a large seed)	4	73.5	.289	
With 90 percent + *Digitaria* and no sorghum	15	70.00	.717	$F = 6.0332$ ($P < .01$ with 1, 17 d.f.)

Data from Quay (1947); wing lengths measured by myself from skins.

TABLE 13B. Field sparrow wing lengths and habitats

Breeding Season

Year	Habitat (small seeds)	Sex	n	$\bar{x}$	SE	Habitat (large seeds) *	Sex	n	$\bar{x}$	SE	F
1964	Pine-broomsedge	♂	15	64.99	.58	Pasture	♂	17	66.18	.99	1.82 *NS*
1966	Broomsedge blackberry	♂	10	64.2	.49	Tall weeds (junipers)	♂	8	66.12	.55	6.85 †
		♀	4	62.0	1.22	(includes pastures)	♀	4	64.50	.29	3.95 *NS*

Winter Season

	Small seed habitat (broomsedge)				Mixed seed size habitat				Large seed size habitat *				
Year	n	$\bar{x}$	SD	SE	n	$\bar{x}$	SD	SE	n	$\bar{x}$	SD †	SE	F
1964–65	9	64.33	9.0	1.00	39	65.48	6.2	.39	15	66.07	6.6	.66	1.25 *NS* †
1965–66		(see note below)			23	63.40	6.8	.54	32	65.00	4.8	.38	6.13 †

Note: In 1965–66, savannah sparrows were sparse and field sparrows were found largely in typically savannah sparrow habitats. Almost none could be found in pure broomsedge habitats.

* Large seed size habitats were those more typically occupied by larger sparrows, especially savannah and song sparrows, and juncos. Broomsedge seeds *are* very small, and Quay (pers. comm.) observed that they were typically broken before eaten. But stomach analysis of field sparrows in true "large seed" habitats were not conducted by me. Basically, "large seed habitat" means no broomsedge.

† $p < .01$. SD = standard deviation. NS = no significant difference.

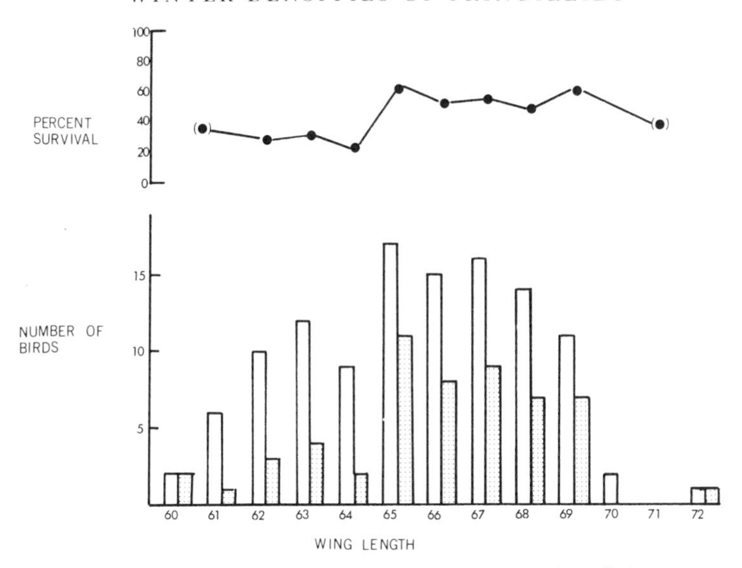

FIGURE 64. Wing length versus survival in a population of wintering field sparrows. The open bar gives number of birds handled for each wing length, and the solid bar gives the number of survivors after a month or more of hard winter. Percentage survival is given above. There is little or no trend visible. Extremely small or large birds all survived.

into neither category and we must at present be vague with regard to what exact body-size distributions will result from a given seed-size distribution.

Availability of Resources

Presuming that we shall eventually understand how to take a distribution of available resource types and predict the distribution of sizes of organisms feeding thereon, it becomes of value to study the factors determining (1) how much of resource is produced and (2) how much of that can be utilized. Of enormous interest would be a study of the size distribution of seeds available to each species and of seeds taken by each species. Study in this area has been started by West (1967). The study by Smith (1970) on coevolution in squirrels and pine cones raises an additional problem. I have treated the seed-size and place distribution as given, and have discussed just the evolution and dynamics of the consumers. But Smith has shown that the resource is evolving as well, with consequences for the consumers. I have not attempted to couple coevolution of resource and consumer with an analysis of population dynamics, but that is clearly what is needed.

SUMMARY

If a population is limited by one season, then the regulation of population by habitat during that season is basic to any understanding of the species' abundance. Considering only those habitats that sustain self-reproducing populations, it should be possible to specify *a priori* the suitability-density curves or the birth-rate or death-rate curves that uniquely determine the distribution and abundance of the species. An initial attempt to solve this problem for populations of wintering sparrows is based on the assumption that food is limiting. Then the density by habitat is determined by (1) how much food is produced, (2) how much is taken by the birds, and (3) how much each bird needs.

A theory of body size is developed to answer the third component, which suggests that sparrows which are food-limited in winter and not limited by breeding territoriality should have body sizes determined by an interaction of seed-size and temperature. A food-limited sparrow in winter should be as small as possible to reduce food needs but large enough to permit sufficiently rapid digestion to meet energy (heat) losses. If a sparrow is climate-limited, a larger body size is optimum, allowing the bird to digest all that it can find. Sparrow males competing for breeding territories are exposed to selection for the largest possible body size. They will increase in size until limited by the ability to find enough food to meet energy (heat) losses in winter. The optimum body size, along with temperature, specifies a sparrow's wintertime energy needs; these needs depend heavily on the particular seed types consumed. Finally, as temperature decreases, the range of permissible sizes diminishes and the species is forced into its special niche.

In analyzing utilization of resources, the problem arises of how an organism of particular body size (and shape) utilizes a continuous distribution of resources. Evidently a certain amount of generalization ability is required, the amount varying with the quality and abundance of the resource. Depending on how much each organism generalizes, there is overlap in utilization between species and organisms. The combined utilization takes up all the available resources. It seems likely that each species has a certain distribution of resources available to it. If its members individually take a narrow range of resources, the species will generate ecotypic variation (through gene frequency dependent selection) in order to utilize all of the resources available.

It is not known exactly how the species distribution of morph types is evolved in response to a given distribution of resource types. Nor is there much evidence or theory indicating what determines availability or production of resources. Thus, the problem of densities of wintering sparrows remains to be solved.

CHAPTER EIGHT

Summary

I began these studies with the general aim of understanding what regulates populations of species. The main theory available when I started was the logistic model and its various competition and predation derivatives as developed, for example, by Slobodkin (1961). This theory, however, seemed to have specific applications only in the laboratory with bottles of bacteria, protozoans, or flour beetles. It did not seem applicable to vertebrate populations at all (Lack, 1966, in his summary of recent research on populations of birds mentions this theory only in a general discussion, never in reference to actual analyses of data or specific cases). In fact, the theory seemed little used in nature even for the short-generation species that fit its assumptions so well.

I supposed, on encountering these difficulties, that a different theoretical approach was needed and eventually decided that the main weakness of the logistic models was the lack of seasonality in them. Most populations live in a seasonal environment. My thesis is that this makes a difference.

The primary part of my analysis is the development of seasonal models, and this was accomplished in Chapters 1 and 2. Having these models at hand, I applied them directly to field data (Chapter 3) and to some other theoretical ideas (Chapter 4). The point of view contained in the models has some indirect value as well, and I attempted to demonstrate this in Section II. The evident dispersive or mobile nature of seasonal populations led to a theory of habitat distribution (Chapter 5). An application of this theory, taking all seasonal effects into account is provided in Chapter 6, where the breeding distribution of field sparrows and other small passerines is analyzed. Some more specialized theoretical extensions are also included. In Chapter 7 the possibility of a species being limited in just the nonbreeding season was used in proposing a theoretical structure for understanding, habitat by habitat, the size of fringillid (sparrow) populations. In this theory, the breeding activities of the species are not considered, demonstrating their possible irrelevance in control of population numbers.

This monograph concentrates too much on birds. Perhaps the theoretical ideas have little utility in other taxonomic groups (plant, insect, etc.), but I optimistically think otherwise. I have endeavored to present the theories generally, sometimes succeeding, sometimes not, and have presented the bird studies as an example of application. I hope that the reader familiar with other taxonomic groups will be able to extract some general ideas from these specific applications.

CONCLUSION

One can suppose that the purpose of population ecology or population biology is to satisfy our separate curiosities about the nature and form of life about us. Or, as Henry Horn has expressed it, to answer the questions of children when we walk with them through a wood. I think, however, that there is a further objective: to find systems in nature analogous to those we work under ourselves. I am quite uncertain about the real value of such analogues; perhaps finding our own behaviors and ecologies more or less represented in "lower" organisms makes us better able to view our own place in nature objectively. Perhaps, too, analogous systems can be used for experimental work.

SUMMARY

Looking back over the research I have presented, I wonder what, if anything, can be made of the various analogies that stand out? For example, there seemed to be, as Wynne-Edwards (1962) might have predicted, "subdominant" populations of field sparrows which are relegated to habitats that are inferior as far as winter survival is concerned (and it seems to be winter food that is limiting). These populations are apparently kept in their place by social discrimination, and have high birth and death rates. Thus, these fringe populations are in certain ways analogous to human ghetto populations.

Again, an immature field sparrow evidently must somehow learn which of several available niches it must select for maximum winter survival. If it is smaller than average, it must learn to select habitats with smaller seeds; if its legs are shorter, it must learn to go where perching while feeding is necessary. In short, it must effect learning analogous to the "search for identity" problem of human young people.

These analogies strike me as a bit extreme; however, I wonder if they would seem so to an impartial observer who has the mental freedom to say "man is another animal" and mean it. Perhaps there is nothing to be gained by making these observations. On the other hand, perhaps man's economic and social life can be subjected to the same kind of comparative analysis that his physiology now receives.

References

Allee, W. C., A. E. Emerson, O. Park, T. Park, and K. P. Schmidt. 1949. *Principles of Animal Ecology.* W. B. Saunders, Philadelphia. xii + 837 pp.

Andrewartha, H. G., and L. C. Birch. 1954. *The Distribution and Abundance of Animals.* Univ. Chicago Press, Chicago. 782 pp.

Ardrey, R. 1967. *The Territorial Imperative.* Dell Publishing Co., New York. 390 pp.

Audubon Field Notes. 1947–1967. Natl. Audubon Society, New York.

Brown, J. L. 1964. The evolution of diversity in avian territorial systems. *Wilson Bull.* 76:160–169.

Brown, J. L. 1969a. Territorial behavior and population regulation in birds. *Wilson Bull.* 81:293–329.

Brown, J. L. 1969b. The buffer effect and productivity in tit populations. *Amer. Nat.* 103:347–354.

Clark, L. R., P. W. Geier, R. D. Hughes, and R. F. Morris. 1967. *The Ecology of Insect Populations.* Methuen & Co. Ltd., London. xiii + 232 pp.

Cody, M. L. 1971. Finch flocks in the Mohave Desert. *Theor. Pop. Biol.* 2:142–158.

Cox, G. W. Unpublished ms. The relation of frequency of migration to species diversity in bird communities.

Cox, G. W. 1968. The role of competition in the evolution of migration. *Evolution* 22(1):180–192.

Crowell, K. L. 1962. Reduced interspecific competition among the birds of Bermuda. *Ecology* 43:78–88.

Davidson, J., and H. G. Andrewartha. 1947a. Annual trends in a natural population of *Thrips imaginis* (Thysanoptera). *J. Anim. Ecol.* 17:193–199.

Davidson, J., and H. G. Andrewartha. 1947b. The influences of rainfall evaporation and atmospheric temperature on fluctuations in the size of a natural population of *Thrips imaginis* (Thysanoptera). *J. Anim. Ecol.* 17:200–222.

Eberhardt, L. L. 1970. Correlation, regression and density dependence. *Ecology* 51:306–310.

Emlen, J. M. 1966. The role of time and energy in food preference. *Amer. Nat.* 100:611–617.

Engels, W. L. 1940. Structural adaptations in Thrashers (mimidae; genus *Toxatoma*) with comments on interspecific relationships. *Univ. Calif. Publ. Zool.* 42:341–400.

Errington, P. L. 1945. Some contributions of a fifteen year local study of the northern bobwhite to a knowledge of population phenomena. *Ecol. Monographs* 15:2–34.

Errington, P. L. 1957. Of population cycles and unknowns. *Cold Spring Harbor Symp. Quant. Biol.* 22:287–300.

Fretwell, S. D. 1968. Habitat distribution and survival in the field sparrow (*Spizella pusilla*). *Bird Banding* 34(4):293–306.

Fretwell, S. D. 1969a. Dominance behavior and winter habitat distribution in juncos (*Junco hyemalis*). *Bird Banding* 40:1–25.

Fretwell, S. D. 1969b. Ecotypic variation in the non-breeding season in migratory populations: a study of tarsal length in some Fringillidae. *Evolution* 23(3)406–420.

Fretwell, S. D. 1970. On territorial behavior and other factors influencing habitat distribution in birds. III. Breeding success in a local population of field sparrows. *Acta Biotheoretica* XIX(1):45–52.

Fretwell, S. D., and J. S. Calver. 1970. On territorial behavior and other factors influencing habitat distribution in birds. II. Sex ratio variation in the Dickcissel. *Acta Biotheoretica* XIX:37–44.

Fretwell, S. D., and H. L. Lucas, Jr. 1970. On territorial behavior and other factors influencing habitat distribution in birds. I. Theoretical development. *Acta Biotheoretica* XIX(1):16–36.

Ghiselin, M. T. 1966. On psychologism in the logic of taxonomic controversies. *Sys. Zool.* 15:207–215.

Gibb, J. 1961. Bird populations. In A. J. Marshall (ed.). *Biology and Comparative Physiology of Birds.* II. Academic Press, New York. pp. 413–446.

Grant, P. R. 1966. Further information on the relative length of the tarsus in land birds. *Yale Peabody Museum Nat. Hist. Postilla No.* 98:1–13.

Harris, V. A. 1964. *The Life of the Rainbow Lizard.* Hutchinson Tropical Monographs. Hutchinson & Co., London. 174 pp.

Helms, C. W. 1968. Food, fat and feathers. *Amer. Zool.* 8:151–167.

Hinde, R. A. 1952. *Behavior of the Great Tit (Parus major) and Some Other Related Species.* E. J. Brill, Leiden. 201 pp.

Holling, C. S. 1966. The functional response of invertebrate predators to prey density. *Men. Ent. Soc. Canada* 48:86 pp.

Howard, R. 1920. *Territory in Bird Life.* Atheneum, London. 239 pp.

Hughes, R. D. 1963. Population dynamics of the cabbage aphid, *Brevicoryne brassicae* (L.). *J. Anim. Ecol.* 32:393–424.

Huxley, J. S. 1934. A natural experiment on the territorial instinct. *Brit. Birds* 27:270–277.

Johnston, R. F. 1961. Population movements of birds. *Condor* 63:386–388.

Kendeigh, S. C. 1969. Tolerance of cold and Bergmann's rule. *Auk.* 86:13–25.

Kiritami, K., N. Hokyo, T. Sasaba, and F. Nakasuji. 1970. Studies on population dynamics of the green leafhopper, *Nephotettix cincticeps* Uhler: regulatory mechanism of the population density. *Res. Pop. Ecology* 12:137–153.

Kluyver, H. N. 1951. The population ecology of the great tit, *Parus m. major. L. Ardea* 39:1–135.

Kluyver, H. N., and L. Tinbergen. 1953. Territory and regulation of density in Titmice. *Arch. Neerl. Zool.* 10:265–289.

Krebs, J. R. 1971. Territory and breeding density in the Great Tit, *Parus major* L. *Ecology* 52:2–22.

Kuno, E., and N. Hokyo. 1970. Comparative analyses of the population dynamics of rice leafhoppers, *Nephotettix cincticeps* Uhler, and *Nilaparvata lugens* Stal, with special reference to the natural regulation of their numbers. *Res. Pop. Ecology* 12:154–184.

Lack, D. 1954. *The Natural Regulation of Animal Numbers.* Oxford Univ. Press, London. 343 pp.

Lack, D. 1964. A long term study of the great tit (*Parus major*). *J. Anim. Ecol.* 33(suppl):159–173.

Lack, D. 1966. *Population Studies of Birds.* Clarendon Press, Oxford. 341 pp.

Levins, R. 1962. Theory of fitness in a heterogeneous environment. I. The fitness set and adaptive function. *Amer. Nat.* 96:361–373.

Levins, R. 1966. The strategy of model building in ecology. *Amer. Sci.* 54:421–431.

MacArthur, R. H. 1957. On the relative abundance of bird species. *Proc. Natl. Acad. Sci. U.S.* 43:293–295.

MacArthur, R. H. 1958. Population ecology of some warblers of northeastern coniferous forests. *Ecology* 39:598–619.

MacArthur, R. H. 1959. On the breeding distribution of North American migrants. *Auk.* 76:218–225.

MacArthur, R. H. 1965. Patterns of species diversity. *Biol. Rev.* 40:510–533.

MacArthur, R. H., and R. Levins. 1964. Competition, habitat selection and character displacement in a patchy environment. *Proc. Natl. Acad. Sci. U.S.* 51: 1207–1210.

MacArthur, R. H., and R. Levins. 1967. The limiting similarity, convergence, and divergence of coexisting species. *Amer. Nat.* 101:377–385.

MacArthur, R. H., and J. W. MacArthur. 1961. On bird species diversity. *Ecol.* 42(3):594–598.

MacArthur, R. H., and E. Pianka. 1966. On optimal use of a patchy environment. *Amer. Nat.* 100:603–609.

MacArthur, R. H., and E. O. Wilson. 1963. An equilibrium theory of insular zoogeography. *Evolution* 17:373–387.

MacArthur, R. H. and E. O. Wilson. 1967. *The Theory of Island Biogeography.* Monographs in Population Biology. Princeton Univ. Press. 203 pp.

Mayfield, H. 1961. Nesting success calculated from exposure. *Wilson Bull.* 73:255–261.

Newton, J. 1967. The adaptive radiation and feeding ecology of some British finches. *Ibis* 109:33–98.

Nice, M. M. 1937. Studies in the Life History of the Song Sparrow. I. *Trans. Linn. Soc. N.Y.* 4:viii + 246 pp.

Nice, M. M. 1957. Nesting success in altricial birds. *Auk.* 74:305–321.

Nicholson, A. J. 1957. The self adjustment of population to change. *Cold Spring Harbor Symp. Quant. Biol.* 22:153–173.

Noble, G. K. 1939. The role of dominance in the social life of birds. *Auk.* 56:263–272.

Nolan, V., Jr. 1963. Reproductive success of birds in a deciduous scrub habitat. *Ecology* 44:305–313.

Oosting, H. J. 1942. An ecological analysis of the plant communities of Piedmont, North Carolina. *Am. Midl. Nat.* 28:1–126.

Orians, G. H. 1962. Natural selection and ecological theory. *Amer. Nat.* 96:257–263.

Orians, G. H. 1969. On the evolution of mating systems in birds and mammals. *Amer. Nat.* 103:589–603.

Orians, G. H., and H. Horn. 1969. Overlap in foods and foraging of four species of blackbirds. *Ecology* 50:930–938.

Osterhaus, Sister M. Benita. 1962. Adaptive modifications in the leg structure of some North American warblers. *Amer. Midl. Natur.* 47:474–486.

Peterson, R. G., H. L. Lucas, and G. O. Mott. 1965. Relationship between rate of stocking and per animal and per acre performance on pasture. *Agronomy J.* 57:27–30.

Pianka, E. R. 1966. Latitudinal gradients in species diversity: a review of concepts. *Amer. Nat.* 100:33–46.

Pielou, E. C. 1969. *An Introduction to Mathematical Ecology.* John Wiley & Sons, New York. 286 pp.

Quay, T. L. 1940. The ecological succession of winter birds at Raleigh, North Carolina. MS thesis. North Carolina State College Library. 71 pp.

Quay, T. L. 1947. Winter birds of upland plant communities. *Auk.* 64:382–388.

Quay, T. L. 1957. The Savannah Sparrow in winter in the lower piedmont of North Carolina. *J. Elisha Mitchell Sci. Soc.* 73:378–388.

Ricklefs, R. E. 1969. An analysis of nesting mortality in birds. *Smithsonian Cont. Zool. 9,* 48 pp.

Rosenzweig, M. L., and R. H. MacArthur. 1963. Graphical representation and stability conditions of predator interactions. *Amer. Nat.* 97:209–223.

Sabine, W. S. 1955. The winter society of the Oregon Junco; the flock. *Condor* 57:88–111.

Salisbury, E. J. 1942. The reproductive capacity of plants, studies in quantitative biology. *London Bull.:* 244 pp.

Selander, R. K. 1965. On mating systems and sexual selection. *Amer. Nat.* 49:129–141.

Selander, R. K. 1966. Sexual dimorphism and differential niche utilization in birds. *Condor* 68:113–151.

Skutch, A. F. 1949. Do tropical birds rear as many young as they can nourish? *Ibis* 91:430–455.

Skutch, A. F. 1954. Life Histories of Central American Birds. *Pacific Coast Avifauna* 31:1–448.

Slobodkin, L. B. 1961. *Growth and Regulation of Animal Populations.* Holt, Rinehard, and Winston, New York. 184 pp.

Smith, C. C. 1970. The coevolution of pine squirrels (Tamiasciurus) and conifers. *Ecol. Monographs* 40:349–371.

Smith, F. E. 1961. Density dependence in the Australian thrips. *Ecology* 42:403–407.

Smith, J. M. 1964. Group selection and kin selection. *Nature, London* 201:1145–1147.

Steel, R. G., and J. H. Torrie. 1960. *Principles and Procedures of Statistics.* McGraw-Hill, New York. 481 pp.

Takasaki, R. 1964. Reproductive curve with two equilibrium points: a consideration on the fluctuation of insect population. *Res. Popul. Ecol.* 6:28–36.

Thoday, J. M. 1959. Effects of disruptive selection. I. Genetic flexibility. *Heredity* 13:187–218.

Thompson, D'Arcy. 1917. *On Growth and Form.* Cambridge Univ. Press, Cambridge. 1116 pp.

Tinbergen, L. 1960. The dynamics of insect and bird populations in pine woods. *Arch. Neerl. Zool.* 13:259–472.

Tinkle, D. W. 1967. The life and demography of the side-blotched lizard, *Uta stansburiana.* Misc. Publications. Museum of Zoology. Univ. Michigan No. 132. 182 pp.

Tricker, R. A. R. 1965. *The Assessment of Scientific Speculation.* American Elsevier, New York. 200 pp.

Udvardy, M. D. F. 1957. An evaluation of quantitative studies in birds. *Cold Spring Harbor Symp. Quant. Biol.* 22:301–311.

West, G. C. 1967. Nutrition of the tree sparrows during winter in central Illinois. *Ecology* 48:58–66.

Wynne-Edwards, V. C. 1962. *Animal Disperson in Relation to Social Behavior.* Hafner Publishing Co., New York. 653 pp.

Zaret, T. M., and A. S. Rand. 1971. Competition in tropical stream fishes: support for the competitive exclusion principle. *Ecology* 52:336–342.

Zimmerman, J. L. 1965. Bioenergetics of the Dickcissel, *Spiza americana. Physiol. Zool.* 38:370–389.

Zimmerman, J. L. ms. Selection for nesting time in the dickcissel: effect of cowbird parasitism and predators.

Zimmerman, J. L. 1971. The territory and its density dependent effect in *Spiza americana. Auk.* 88:591–612.

Index